R.L. Timings
C.ENG., F.I.P.C., M.I.Pro...

Materials technology
Level 2

Longman London and New York

Longman Group Limited
Longman House, Burnt Mill, Harlow
Essex CM20 2JE, England
Associated companies throughout the world

*Published in the United States of America
by Longman Inc., New York*

© Longman Group Limited 1984

All rights reserved; no part of this publication may be
reproduced, stored in a retrieval system, or transmitted
in any form or by any means, electronic, mechanical,
photocopying, recording, or otherwise, without the
prior written permission of the Publishers.

First published 1984

British Library Cataloguing in Publication Data

Timings, R.L.
  Materials technology, level 2.—(Longman
  technician series. Mechanical and
  production engineering)
  1. Materials
  I. Title
  620.1'1   TA403
  ISBN 0-582-41339-7

Printed in Great Britain
by The Pitman Press Ltd., Bath

General Editors – Mechanical and Production Engineering

H. G. Davies

Vice Principal and Head of Department of Science, Carmarthen Technical and Agricultural College

G. A. Hicks

Lecturer in the Department of Engineering, Carmarthen Technical and Agricultural College

Books published in this sector of the series:

Workshop processes and materials   *R. L. Timings*
Manufacturing technology Level 2   *R. L. Timings*
Manufacturing technology Level 3   *R. L. Timings*
Engineering science for mechanical technicians Level 2   *J. O. Bird and A. J. C. May*
Engineering science for mechanical technicians Level 3   *J. O. Bird and A. J. C. May*
Engineering drawing for technician engineers   *J. D. Poole*
Engineering science Level 1   *D. R. Browning and I. McKenzie Smith*
Motor vehicle engineering drawing for technicians Level 1   *S. J. Zammit*
Motor vehicle engineering drawing for technicians Level 2   *S. J. Zammit*
Motor vehicle engineering science for technicians Level 2   *S. J. Zammit*

Contents

Preface vii

Acknowledgements viii

Chapter 1	**The structure of metals** 1	
Chapter 2	**Binary equilibrium diagrams** 16	
Chapter 3	**Plain carbon steels** 36	
Chapter 4	**The heat treatment of plain carbon steels** 55	
Chapter 5	**Heat-treatment equipment and processes** 75	
Chapter 6	**Common cast irons** 100	
Chapter 7	**Non-ferrous metals and alloys** 114	
Chapter 8	**Polymers** 145	

Index 170

Preface

Materials technology: Level 2 is the first in a series of books concerned with engineering materials to be published in the Longman Technician Series. It has been written to satisfy the requirements of students following the Business & Technician Education Council's (B/TEC) standard unit U80/738. This unit covers materials technology at level 2 in the mechanical and production engineering programme A5.

Materials technology is now an essential unit in the Business & Technician Education Council's model programme. The syllabus items of materials and heat treatment have now been withdrawn from the latest units for manufacturing technology, and have been expanded into units in their own right. Therefore, *Materials technology: Level 2* should be read in conjunction with the revised edition of *Manufacturing technology: Level 2* (which has been updated accordingly) by the same Author.

Subsequent books in this series will cover the requirements of the Technician Education Council's standard units in materials technology at levels 3 and 4.

Finally, I wish to thank: my friends and colleagues who have assisted me in writing this book and checking the proofs; the organisations who have provided up-to-date technical data and illustrations; the publishers and series editors for their help and advice in the preparation of the manuscript; and Mrs Jean Smith for typing the manuscript and supporting documents.

R. L. Timings
1984

Acknowledgements

We are grateful to the following for permission to reproduce copyright material: Edward Arnold (Publishers) Ltd for our Fig. 7.2 from *Metallurgy for Engineers* by E. C. Rollason; BCIRA for our Figs. 6.2, 6.3, 6.4, 6.6, 6.7 and 6.8; Hodder and Stoughton Ltd for our Fig. 1.9 from *Properties of engineering materials* by Higgins: Holt-Saunders Ltd for our Fig. 3.3(b) from *Technician structure and properties of metals* by H. A. Monks and D. C. Rochester.

Chapter 1

The structure of metals

1.1 Introduction

Consider the various metal objects shown in Fig. 1.1. The lathe tailstock is made of cast iron; the spanners are made of steel and the electric cable is made of copper. Each metal is chosen because it has special and individual properties which make it different from the others and suitable for the application shown. But what have these metals got in common? No matter what changes they have undergone during extraction from the ore, refinement, processing, and final manufacture, the form they adopt in Fig. 1.1 is that of a crystalline solid. In fact, all metals are crystalline solids at room temperature, with the notable exception of the metal mercury. In order to assess the behaviour and properties of the many different metals and alloys available to them, so that they can select the most suitable for a given application, the engineer and the metallurgist must understand the crystalline structure of the metal. However, before studying the crystalline structure of metals and alloys it is advisable to revise some basic chemical concepts.

1.2 Atoms

Figure 1.2 (*a*) shows, diagrammatically, an atom of hydrogen (the simplest atom) and Fig. 1.2 (*b*) shows an atom of the metal copper. It can be seen that both atoms consist of a *nucleus* around which are orbiting

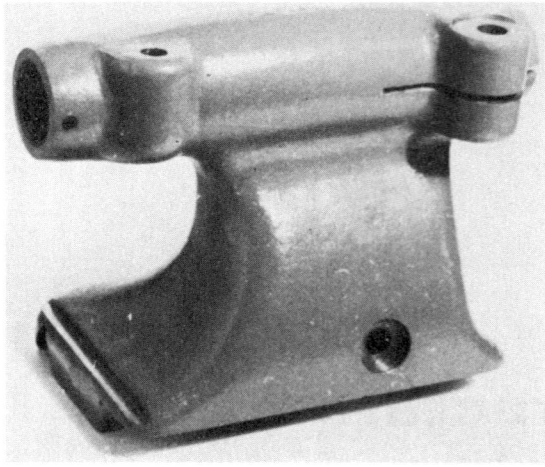

(a)

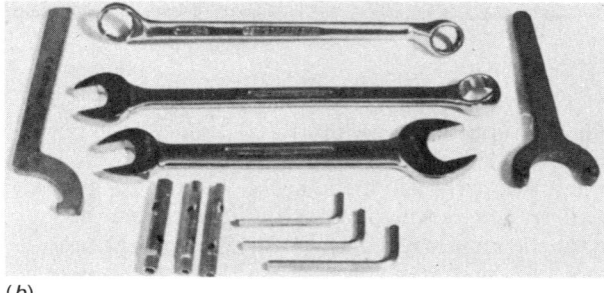

(b)

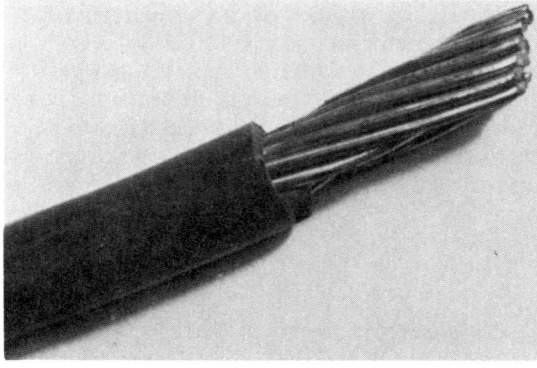

(c)

Fig. 1.1 Typical metal components (a) Lathe tail-stock (b) Spanners and keys (c) Electric cable

one or more *electrons*. Although the electrons spend most of their time in 'shells' as shown in Fig. 1.2 (*b*), these are not as rigid as the drawing implies and the electrons are free to wander anywhere round the nucleus as shown in Fig. 1.2 (*c*). The dots show the location of the electrons at various times. The basic structure of the atom is as follows.

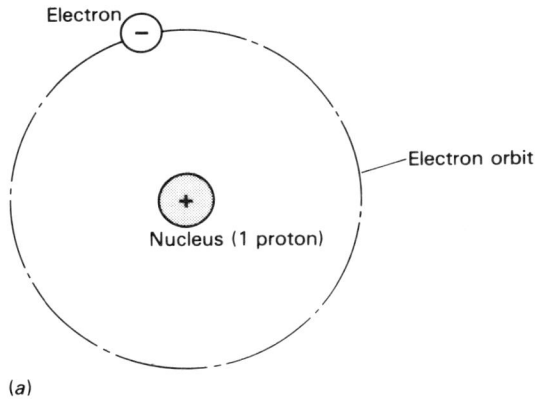

(*a*)

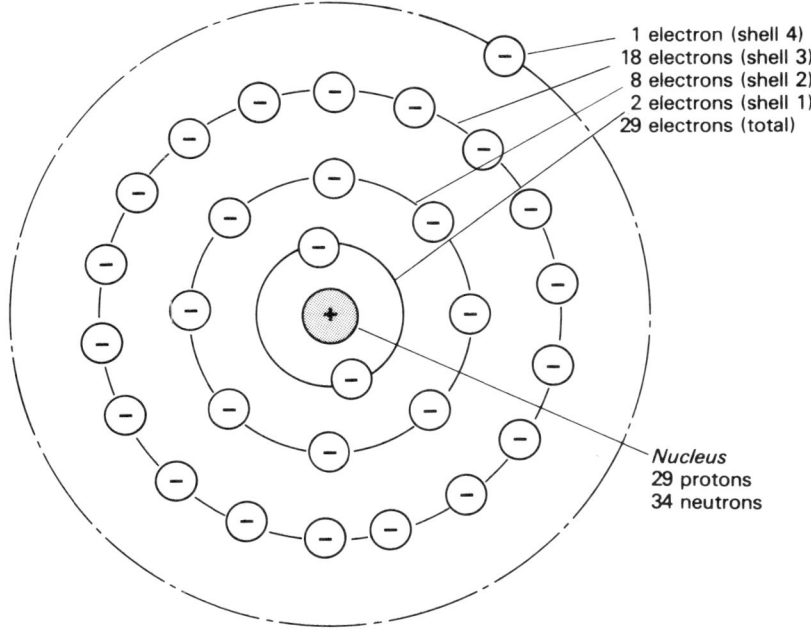

(*b*)

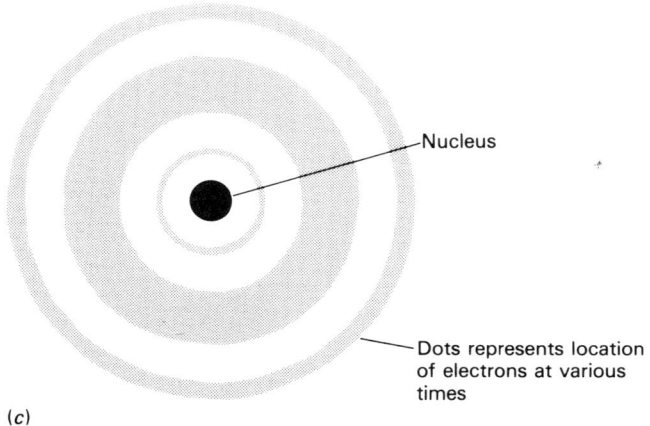
(c)

Fig. 1.2 Typical atomic structures (a) Hydrogen atom (b) Copper atom (c) Movement of electrons round the nucleus

(a) **Nucleus.** This is the basic core of the atom and consists of *protons* and *neutrons*.
(b) **Protons.** These are positively charged particles of very much greater mass than the electrons.
(c) **Neutrons.** These particles have the same mass as protons but carry no electrical charge.
(d) **Electrons.** These particles are negatively charged and orbit the nucleus like planets around a sun. Although electrons are very small and have only 1/1836 the mass of a proton or neutron they are of supreme importance when considering how atoms bond together to form molecules.
(e) **Atoms** can be considered as the smallest particle of a substance which can exhibit all the properties of that substance. An atom is electrically neutral because it has an equal number of electrons and protons. The chemical properties of an atom, that is, how it combines with other atoms, are determined by the number of electrons that it has.
(f) **Ions** are atoms which have gained or lost one or more electrons. Loss of an electron makes the atom electropositive and it is called a *positive ion*. Gaining an electron makes the atom electronegative and it is then called a *negative ion*.

Since positive ions are attracted towards the cathode (negative electrode) of an electrolytic cell, positive ions are also called *cations*. Similarly, negative ions are attracted towards the anode (*positive electrode*) of an electrolytic cell, and are also called *anions*.
(g) **Isotopes.** Since the electron is so small compared with the proton, the mass of the atom can – for all practical purposes – be considered as

concentrated in the nucleus. As has already been stated, the neutron has the same mass as the proton but no electrical charge. Therefore, if the number of neutrons in the nucleus changes, the mass of the atom changes but its chemical properties remain unchanged (although it can influence the radioactivity during nuclear reaction). Atoms which have the same chemical properties but different atomic masses are said to be *isotopic* and are referred to as *isotopes* of a given substance.

1.3 Molecules

So far, the atom has been considered as a single free particle. However, apart from the noble gases such as neon (used in electric discharge tubes for advertising) and argon (used as a gas shield when welding), atoms rarely occur as single particles but are generally associated with other atoms in small or large groups. These groups of atoms are called molecules. Figure 1.3 shows a very simple molecule consisting of two hydrogen atoms. Since they share their electrons they are said to be joined by a *covalent bond*. Some molecules may contain thousands of individual atoms.

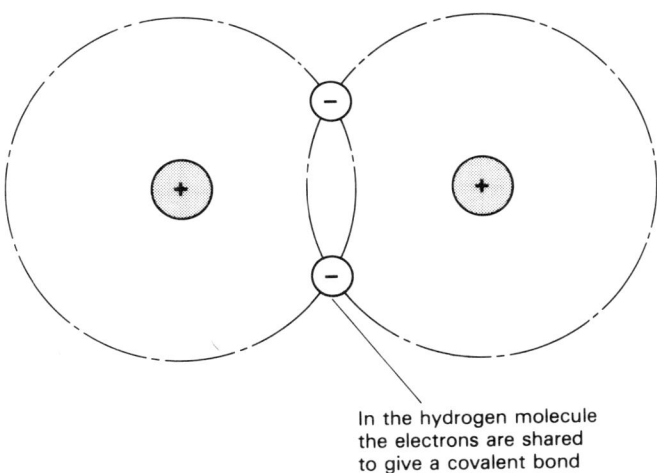

In the hydrogen molecule the electrons are shared to give a covalent bond

Fig. 1.3 The hydrogen molecule

1.4 Elements

A substance composed of atoms all with the same number of electrons is an element. Elements are pure substances incapable of further division and consisting of molecules formed entirely from one type of atom only. For example: iron, carbon, sodium, chlorine and copper. Steel is not an element since it contains both iron and carbon. Table salt is not an element since it consists of both sodium and chlorine (see section 1.6). There are 103 elements at present known to scientists.

1.5 Mixtures

Mixtures are two or more substances in close association but not chemically combined, and which may be separated without recourse to chemical process.

Fig. 1.4 Copper sulphate crystals growing from solution (J. Lewis)

For example, the grit and salt spread on the roads in the winter is a *mixture*. That is, the grit and the salt do not combine chemically and they can be easily separated. If the grit and salt are placed in sufficient water, the salt will dissolve to form a solution and the grit can be filtered off. The salt solution can then be boiled, and as the water evaporates the salt is left behind.

1.6 Compounds

When two or more different types of atoms join together a compound is formed. Generally this compound molecule will have totally different properties to either of the constituent atoms, and a chemical process is necessary to divide the compound into its constituent elements. For example, the metal sodium reacts violently with the poisonous gas chlorine to form the stable salt sodium chloride (common table salt).

Metal alloys are sometimes formed by two metals reacting to form *intermetallic compounds*, but these are relatively rare compared with the number of compounds formed between metals and non-metals, or between non-metals and non-metals.

1.7 Crystals

Many solid substances, including the metals, are *crystalline* in structure. That is, their basic particles are arranged in definite three-dimensional patterns of rigid geometrical form which are repeated many times. Allowed to grow freely, crystals take on a distinctive geometric form as shown in Fig. 1.4.

Non-crystalline 'solids' such as pitch, glass and many 'plastic' materials are said to be *amorphous* (without shape) and are best understood if they are considered as being extremely viscous liquids.

The structure of crystals may be understood if their constituent atoms are considered to be spherical in shape. Figure 1.5 (*a*) shows a simple cubic crystal built up from eight spherical particles.

The dotted lines joining the centres of the spheres represent the *unit cell* of this simple crystal. The unit cell is the geometric figure which illustrates the fundamental grouping of the particles in the solid. To form the crystal this unit cell is repeated many times to form the *space lattice* as shown in Fig. 1.5 (*b*). It is this regular, repetitive pattern of particles which characterises crystalline materials.

All crystal structures can be analysed into fourteen basic lattices. These are called the *Bravais space lattices*. For simplicity only the unit cell of each lattice is shown in Fig. 1.6.

P–type. These are shown in Fig. 1.6 (*a*) and are classified as *primitive*.
C–type. These are shown in Fig. 1.6 (*b*) and are classified as *centred on the 'ab' face*.

I–type. These are shown in Fig. 1.6 (*c*) and are classified as *body-centred*.

F–type. These are shown in Fig. 1.6 (*d*) and are classified as *face-centred*.

R–type. and *H–type* are shown in Fig. 1.6 (*e*) and 1.6 (*f*) respectively.

Of these fourteen possible space lattice formations, only six are met with in metal crystals. Of these, the most common are:

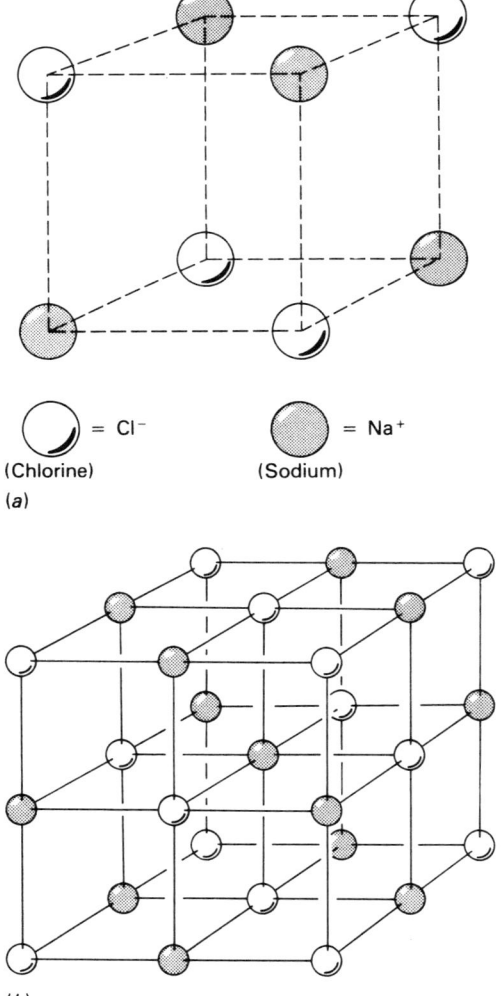

Fig. 1.5 The crystal structure (*a*) Unit cell for sodium chloride (common salt) crystal (*b*) Part of the space lattice for sodium chloride (four unit cells shown)

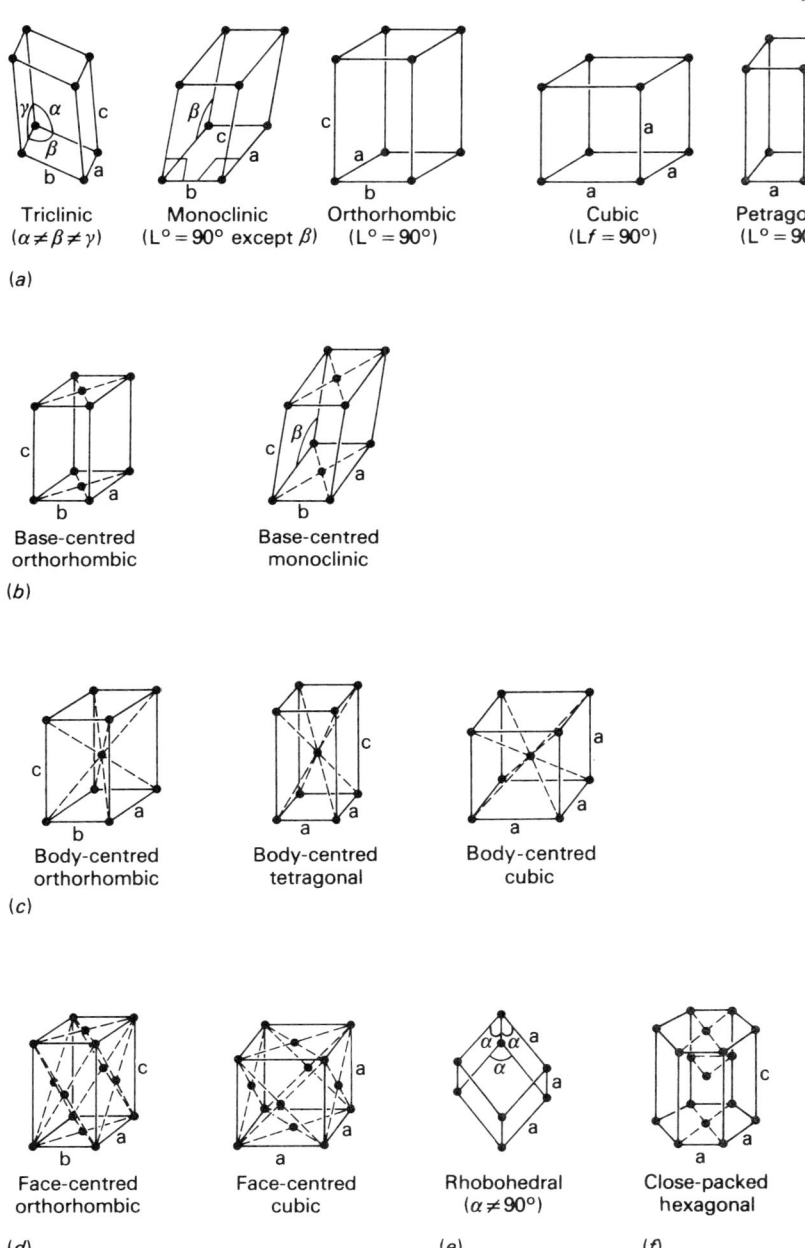

Fig. 1.6 Bravais space lattices (*a*) P-type (primitive space lattices) (*b*) C-type (base-centred on 'ab' face) (*c*) I-type (body-centred) (*d*) F-type (face-centred) (*e*) R-type (*f*) H-type

Body-centred cubic	Face-centred cubic	Close-packed hexagonal
Chromium	Aluminium	Beryllium
Molybdenum	Copper	Cadmium
Niobium	Lead	Magnesium
Tungsten	Nickel	Zinc

The above are only a small number of examples.

1.8 Allotropy

Allotropy is the ability of a substance to exist in more than one physical form. The non-metal carbon is said to be allotropic since it can exist as both diamond and graphite. Both these substances consist solely of carbon atoms, but it is in the crystal structure and the way in which the atoms are bonded together where the difference lies.

The metal iron is another allotropic substance which is why it was not included in the examples given in section 1.6.

Below 910 °C iron has a body-centred cubic space lattice and is referred to as α iron.

Between 910 °C and 1400 °C iron has a face-centred cubic space lattice and is referred to as γ iron.

Above 1400 °C iron has a body-centred cubic space lattice again and is referred to as δ iron.

The allotropy of solids which relies solely on difference in the crystal structure (space lattice) is referred to as *polymorphism*. Many metals are allotropic.

1.9 Grain structure

Although metals are crystalline solids this is not immediately apparent when they are examined under the microscope. Figure 1.7 (*a*) shows the appearance under the microscope of a typical metal specimen which has been polished and etched. Although obviously granular, it is difficult to identify the geometric regularity expected of crystals. This is because crystals can only achieve geometric regularity when they are free to grow without interference. In a metal many crystals commence growth at the same time and eventually collide with each other so that their boundaries are distorted.

The term *grain* is used to describe crystals whose geometric shape has been distorted by contact with adjacent crystals so that their growth is impeded. Figure 1.7 (*b*) shows, diagrammatically, how the atoms within a grain can have the regular geometric pattern expected of a crystal, but how that pattern breaks down at the grain boundary to make way for the geometric pattern of particles in the adjacent grains.

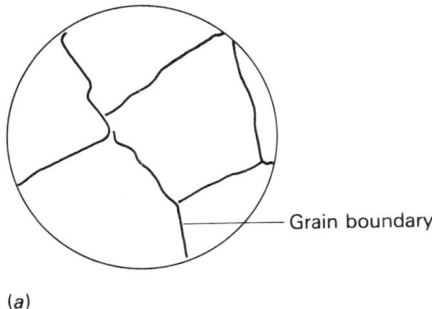

(a)

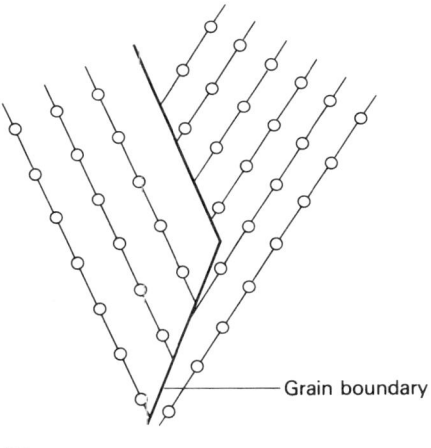

(b)

Fig. 1.7 Grain structure (a) Appearance of granular structure of metal under the microscope after etching (b) Despite the irregular appearance of the grain structure due to boundary interference, the crystal lattice within the grain is correctly ordered

1.10 Crystal growth

All pure elements exist as gases, liquids or solids depending upon the combination of temperature and pressure to which they are exposed at any one time. This is not necessarily true of compounds, many of which break down and decompose at temperatures below their boiling points and some even below their melting points.

In the case of metals, only mercury is a liquid at room temperature, though some metals melt below red heat. A few such as cadmium, mercury and zinc will even boil at low temperatures and can be refined by distillation.

In the liquid state there is no orderly arrangement of atoms in a pure metal and the atoms are free to move with respect to each other, thus a liquid possesses *mobility*.

As the temperature of the molten metal falls, a point is reached where the metal starts to solidify. At this point the atoms change from a disordered or *amorphous* state to an ordered or crystalline state.

Like other pure crystalline substances, pure metals solidify at a fixed single temperature as shown in Fig. 1.8 (*a*). Under industrial conditions the crystal nucleus forms around an impurity particle such as a grain of slag. However, as shown in Fig. 1.8 (*b*), in a very pure metal some under-cooling may occur before nucleation sets in. Amorphous (non-crystalline) solids such as glass, pitch and some 'plastics' exhibit no such change point and, as has been previously stated, these so-called solids are more akin to extremely high-viscosity liquids.

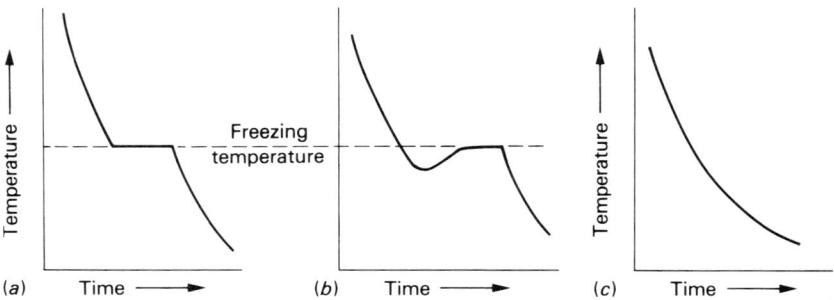

Fig. 1.8 Cooling curves (*a*) Pure metal: no undercooling (*b*) Pure metal: some undercooling (*c*) Amorphous solid: no single freezing temperature

Once the nucleus of the crystal forms it provides a solid/liquid interface where crystallisation can proceed. The nuclei which form will be crystal unit cells, generally face-centred cubic, body-centred cubic or close-packed hexagonal. As the crystal grows on these nuclei it tends to develop spikes and changes into a 'tree-like' shape called a *dendrite*. (Greek *dendron* = a tree). Figure 1.9 shows a typical metal dendrite. The dendritic crystal grows until the spaces between the branches fill up. Growth of the dendrite ceases when the branches of one dendrite meet those of an adjacent dendrite, and eventually the entire liquid solidifies. At this point there is little trace of the dendritic structure left, and it is only possible to see the grains into which the dendrites have grown. The steps in the growth pattern of a crystal from nuclei to grain are shown in Fig. 1.10.

The reason for dendritic growth is as follows. When a liquid at its fusion point (melting point) solidifies it gives up latent heat energy. This is the latent heat of the fusion taken in when the solid was originally melted.

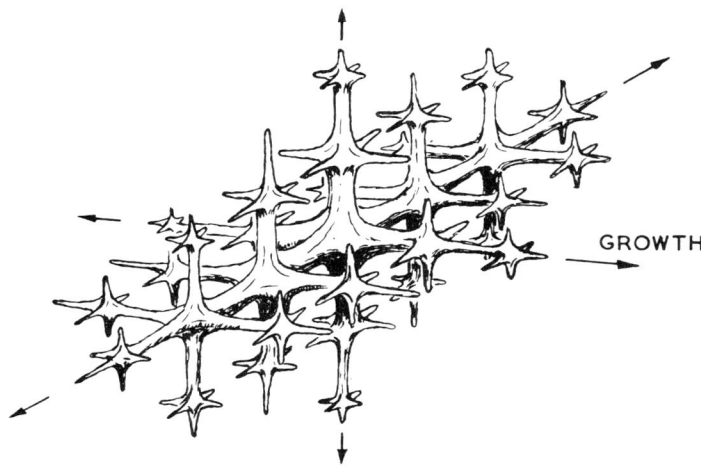

Fig. 1.9 Metallic dendrite growth (R. A. Higgins)

When a solid is heated, the atoms vibrate about fixed points, called lattice points, each atom being held in place by forces of attraction. As the temperature increases, the energy of vibration of each atom increases. At a certain point (temperature) they can overcome the forces binding them to the lattice points, and can escape from their fixed positions. Thus the structure begins to lose its rigidity, that is, it begins to melt. In moving from their fixed positions the atoms or molecules do work against the binding forces, and also as a result energy is used up. This energy is replaced by the heat source of the furnace. Thus the heat energy supplied is used to produce fusion (melting) rather than causing a rise in temperature.

The energy that is used as work to cause the change of state from solid to liquid is called the *latent heat* of fusion. The amount of latent heat energy required to melt 1 kg of a substance is called the *specific latent heat*.

Thus the metal/liquid interface is warmed up by the release of the latent heat energy as solidification occurs. This slows or stops further solidification occurring in that direction. The result of this action is for spikes to develop into regions where the liquid is coolest. As these spikes warm up in turn, due to the release of latent heat energy, forward growth of the crystal is again retarded and secondary and even tertiary spikes are formed.

Although it is hard to relate a dendrite to the well-ordered crystal structures previously considered, it must be remembered that the unit cells and space lattices are very, very small even when compared in size with the dendritic spike. Thus during solidification and crystallisation the ordered pattern of the space lattice is still being built up, but the rate of growth is not uniform in all directions.

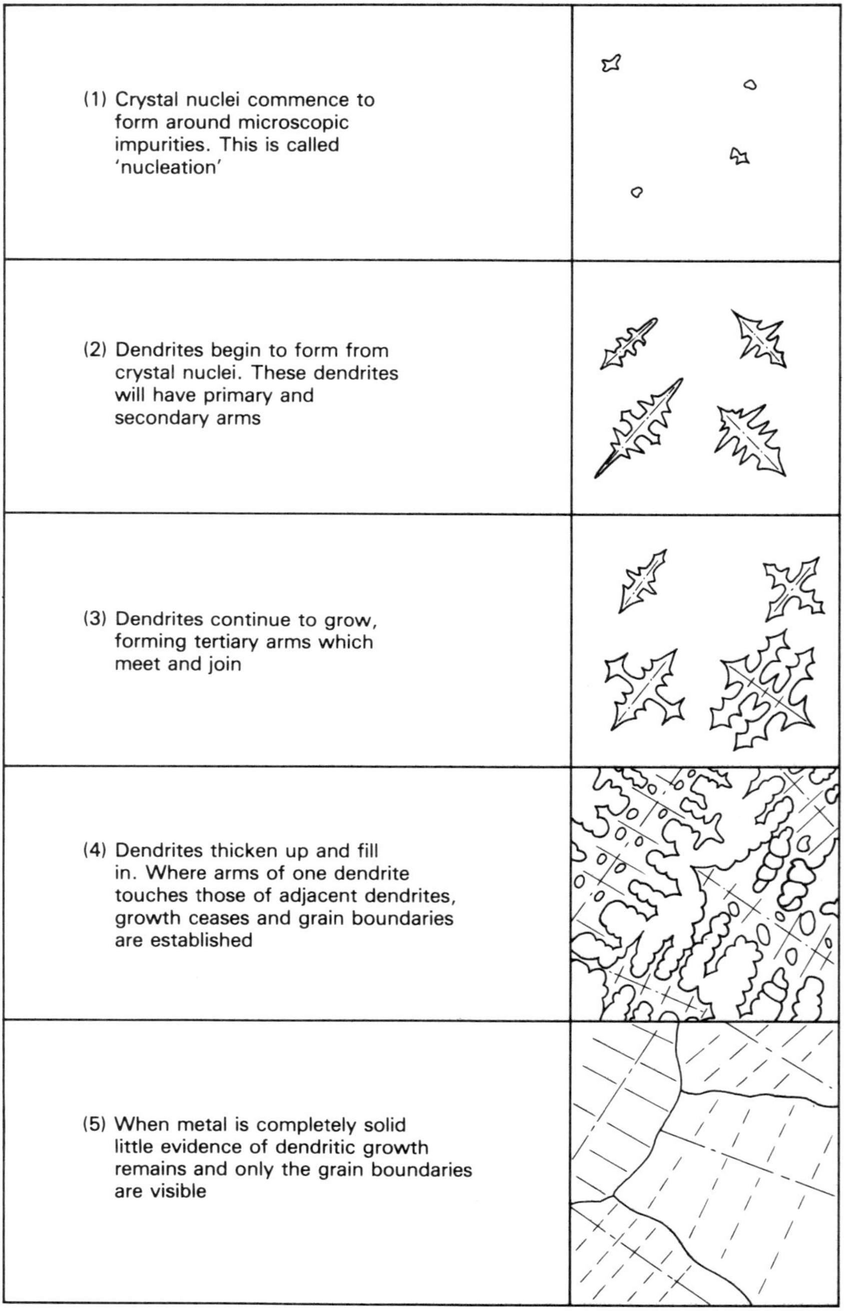

(1) Crystal nuclei commence to form around microscopic impurities. This is called 'nucleation'	
(2) Dendrites begin to form from crystal nuclei. These dendrites will have primary and secondary arms	
(3) Dendrites continue to grow, forming tertiary arms which meet and join	
(4) Dendrites thicken up and fill in. Where arms of one dendrite touches those of adjacent dendrites, growth ceases and grain boundaries are established	
(5) When metal is completely solid little evidence of dendritic growth remains and only the grain boundaries are visible	

Fig. 1.10 Crystal growth

Problems

Section A
1. Describe briefly the essential differences between an element, a compound, and a mixture.
2. State the essential differences between a crystalline and an amorphous solid.
3. State the differences between a 'unit cell' and a 'space lattice'.
4. With the aid of neat sketches show the difference between the unit cell for a:
   (i) body-centred cubic crystal, (BCC);
   (ii) face-centred cubic crystal, (FCC);
   (iii) close-packed hexagonal crystal. (CPH).
5. Explain briefly what is meant by the term 'allotropy'.

Section B
6. Sketch a typical representation of an atom and describe the particles which are to be found in the nucleus and the 'shells' surrounding the nucleus.
7. (a) Explain what changes must occur in an atom if it is to become:
   (i) an anion;   (ii) a cation.
   (b) Explain what is meant when a material is said to be 'isotopic', and explain what changes take place in the atomic structure to bring about this state.
8. (a) Explain in detail the mechanism of crystal growth from the molten liquid to the solid state for a metal.
   (b) With the aid of sketches explain what is meant by the term 'dendrite', and explain how dendritic growth occurs.
9. Explain the essential differences between the crystal structure of a metal and its irregularly shaped 'grain'.
10. Explain in detail what is meant by the term 'latent heat of fusion' and why the temperature of a metal remains constant whilst fusion occurs.

Chapter 2

Binary equilibrium diagrams

2.1 Alloys

A number of metal objects were shown in Fig. 1.1. The conductors in the electric cables are made from copper because of its high electrical conductivity. The connecting rod is made from an alloy steel to give it great strength. The lathe tail-stock is made from cast iron. This is an iron-carbon alloy which melts relatively easily and can be cast into complex shapes.

Pure metals are used where good electrical conductivity, good thermal conductivity good corrosion resistance or all of these properties are required. Since pure metals usually lack the strength required for structural materials, alloys are designed to give superior mechanical properties and they can be 'tailored' to suit a particular application.

An *alloy* is an intimate association of two or more component materials which form a metallic liquid or solid. The component materials may be metal elements, or they may be metal and non-metal elements. They may also be metal elements and chemical compounds.

Useful alloys can only be produced from component materials which are soluble in each other in the molten state. That is, they are completely *miscible*. It would be useless to try and form an alloy from lead and zinc. The molten zinc would float on the molten lead and, on cooling, they would form two separate layers in the solid state with only tenuous bonding at the interface. Alloys are formed in one of three ways:

1. If the alloying components in the molten solution have similar chemical properties, and their atoms are similar in size, they will not

react together but will form solid solutions upon cooling.
2. If the alloying components in the molten solution have different chemical properties they may attract each other and form chemical compounds. These are often referred to as *intermetallic compounds*. Upon cooling the crystals will consist of a mixture of such compounds.
3. In a situation where atoms with different chemical properties attract each other less than those with similar chemical properties, then both intermetallic compounds and solid solutions will be present at the same time. Upon cooling they will tend to separate out at the grain boundaries to form a heterogeneous mixture.

In any alloy the metal which is present in the larger proportion is referred to as the *parent metal* or *solvent*, whilst the metal (or non-metal) present in the smaller proportion is known as the *alloying component* or *solute*.

2.2 Solubility

In order to understand the formation of alloys, it is first necessary to understand the basic principles of solubility in the liquid and solid states.

Sodium chloride (common table salt) dissolves readily in water. In cold water, at room temperature, approximately 35 g of sodium chloride will dissolve in 100 g of water. The exact amount will depend upon the temperature of the water. If more sodium chloride is added to the solution it will not dissolve because the solution is *saturated*. The excess salt will remain as a solid residue. The solubility of the sodium chloride increases only slightly as the temperature of the water increases.

In this example the sodium chloride is dissolved in water, thus:

1. The water is called the *solvent*.
2. The sodium chloride is called the *solute*.
3. The resulting liquid is called the *solution*.

Figure 2.1 shows the difference between complete and partial *solubility*. Copper sulphate can also be dissolved in water but, unlike sodium chloride, its solubility increases substantially as the temperature of the solvent increases. This is shown in Fig. 2.2.

Consider 50 g of copper sulphate being dissolved in 100 g of water as shown by the broken line.

(a) Above 80 °C the water is capable of dissolving more than 50 g of copper sulphate, so the solution is said to be *unsaturated*.
(b) At 80 °C the water will dissolve a maximum of 50 g of copper sulphate, so the solution is said to be *saturated*.
(c) Below 80 °C the water dissolves less than 50 g of copper sulphate. For example, at 40 °C only 30 g of copper sulphate can be dissolved in 100 g of water (only 30 g of $CuSO_4$ can be 'held in solution') and

the balance of 20 g of copper sulphate will be precipitated out of solution as a solid residue.

Substances which will not dissolve in a solvent are said to be *insoluble*. *Note:* A substance may be insoluble in one solvent, but soluble in another.

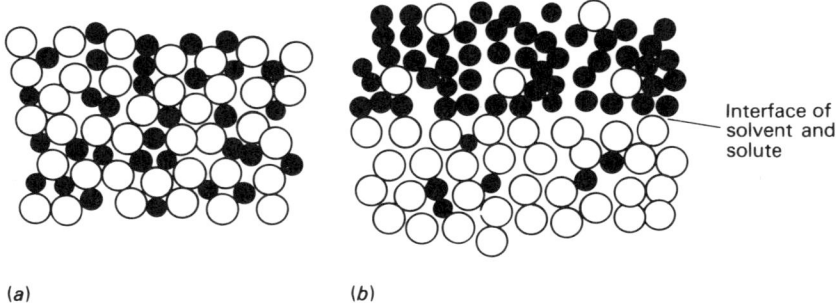

(a) (b)

Fig. 2.1 Solubility (*a*) Complete solubility (*b*) Partial solubility

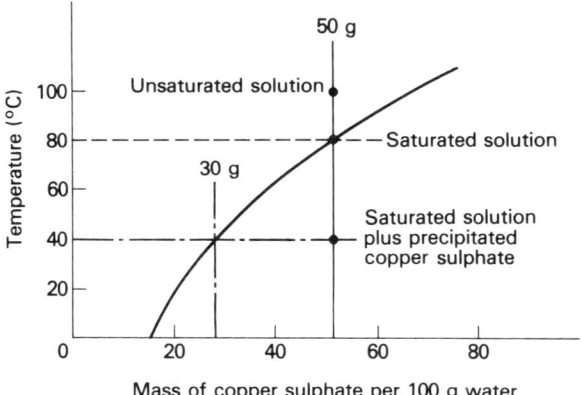

Fig. 2.2 Solubility curve for copper sulphate

2.3 Solid solutions

Most metals are completely and mutually soluble (they are miscible) in the liquid state, that is, when they are molten. Some, such as copper and nickel, not only form solutions in the molten or liquid state, but remain in solution upon cooling and become *solid solutions*. There are two sorts of solid solution:

1. *substitutional* solid solutions;
2. *interstitial* solid solutions.

The copper–nickel alloy mentioned previously is a *substitutional solid solution*. The more important factors governing the formation of a substitutional solid solution are:

(a) *Atomic size*. The atoms of the solvent and solute must be approximately the same size. If the atom diameters vary by more than 15 per cent the formation of a substitutional solid solution is highly unlikely.

(b) *Electrochemical series*. All metals are electropositive to some degree. If there is only a small difference in charge between the alloying components then they will probably form a solid solution. Conversely if their positive charges are very dissimilar they are more likely to form intermetallic compounds.

(c) *Valency*. A metal of lower valency is more likely to dissolve one of higher valency than the other way round, assuming the conditions set out in (a) and (b) are also favourable. This holds good particularly for monovalent metals such as copper, silver and gold.

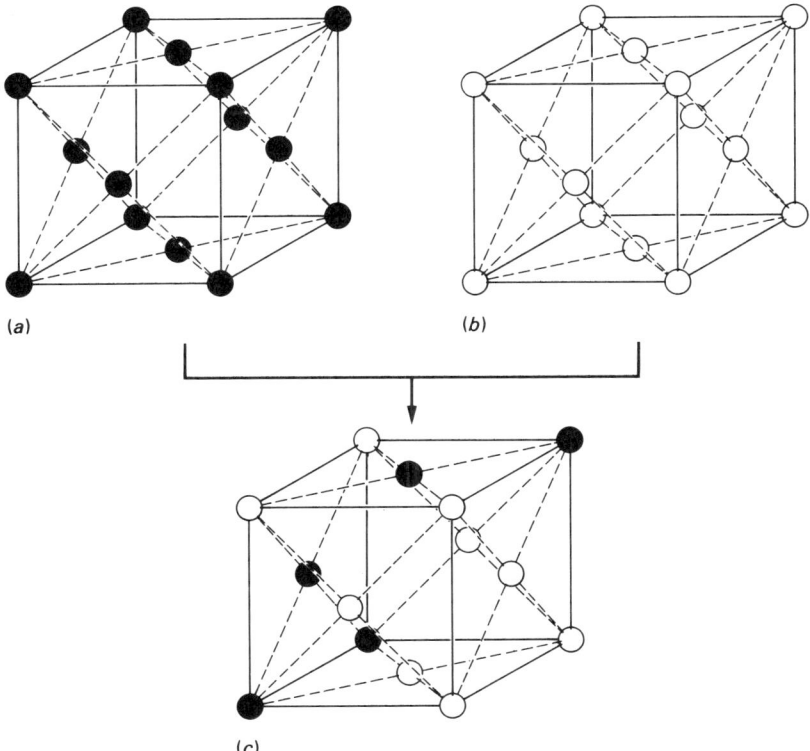

Fig. 2.3 Substitutional solid solution (a) Face-centred cubic crystal of copper (b) Face-centred cubic crystal of nickel (c) Substitutional solid solution of copper and nickel

Figure 2.3 shows that both copper and nickel form face-centred cubic crystals. When these two metals are in solid solution they form a single face-centred cubic lattice with atoms of nickel replacing atoms of copper in the lattice. Hence the term substitutional solid solution.

The substitution can be *ordered*, with the atoms taking up regular fixed positions of geometric symmetry in the lattice. However, most solid solutions are disordered, with the solute atoms appearing virtually at random throughout the solvent lattice.

Interstitial solid solutions are formed when the solute atoms are small enough to lie between the solvent atoms as shown in Fig. 2.4. For example, carbon atoms can form an interstitial solid solution with face-centred cubic crystals of iron.

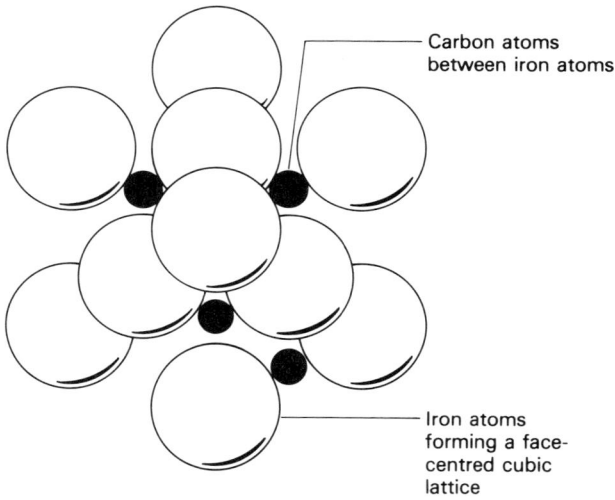

Fig. 2.4 Interstitial solid solution

2.4 Intermetallic compounds

It has already been stated that where the components of the alloy are sufficiently different chemically, they will tend to form compounds rather than solid solutions. In general, intermetallic compounds tend to be hard and brittle and are thus less useful for engineering alloys than the tough and ductile solid solutions. Intermetallic compounds are most widely found in bearing metals where they form hard, wear-resistant pads with a low coefficient of friction, set in a matrix of a tough solid solution.

2.5 Cooling curves

Most substances can exist as gases, liquids and solids, depending upon their temperature. Water is one such substance, which can exist as a gas or vapour (steam) if it is sufficiently hot; as a liquid; and as a solid (ice) if it is sufficiently cold.

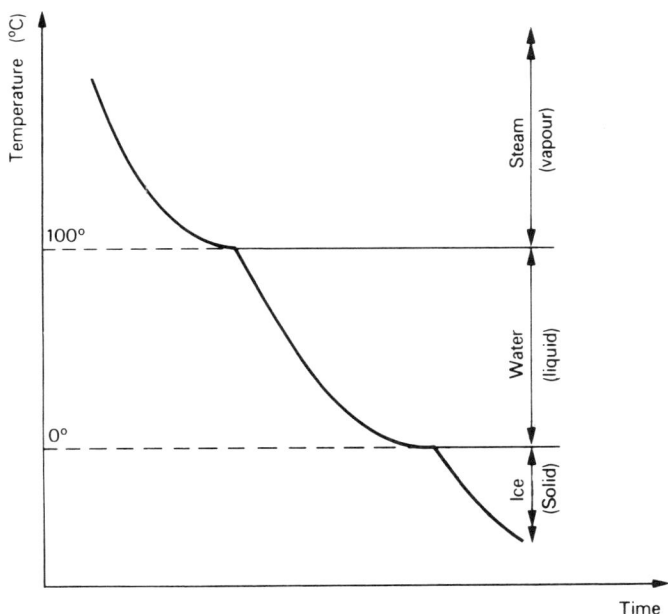

Fig. 2.5 Cooling curve for water

If water is raised to its boiling point and allowed to cool slowly, the change in temperature with time can be plotted as a graph as shown in Fig. 2.5. Such a graph is called a cooling curve. It will be seen that when a change of state occurs (such as liquid water to solid ice) there is a short pause in the cooling process. This pause is referred to as an arrest point and is the result of the substance absorbing or giving out latent heat. Latent heat is the heat energy required to produce a change of state in a substance at a constant temperature. The gaseous, liquid and solid states of a substance are often referred to as phases. A substance is said to be in the gaseous phase, the liquid phase or the solid phase. It will be seen later that phase changes can also occur in solids.

A physical change of state during the cooling, or heating, of a substance is always accompanied by an arrest point in the cooling or heating curve.

The cooling curve shown in Fig. 2.5 is typical of pure substances and applies equally well to any pure metal. Alloys consist of two or more components and, to understand their behaviour on cooling, the above explanation must now be extended to encompass a solution.

A suitable solution is that of domestic table salt (sodium chloride) in water. Figure 2.6 shows the cooling curve for pure water compared with the cooling curve for a salt-water solution for the temperature range covering the liquid and solid phases. It will be seen that the salt-water solution has two arrest points and that both these are below the freezing point of water.

A salt-water solution has a lower freezing point than pure water and at 0 °C no change of state occurs. However, as cooling continues, droplets of pure water separate out from the solution and immediately change into ice particles. This occurs at the upper arrest point, which is not usually too well defined, and the process of separation continues as the temperature of the remaining solution is further reduced. Thus, as the temperature continues to fall, more and more water separates out and freezes, causing the concentration of the remaining salt water to increase. When the lower arrest point is reached, even the concentrated salt-water solution freezes and no liquid phase is left. The solid so formed consists of a mixture of fine crystals of pure water (ice) and fine crystals of salt.

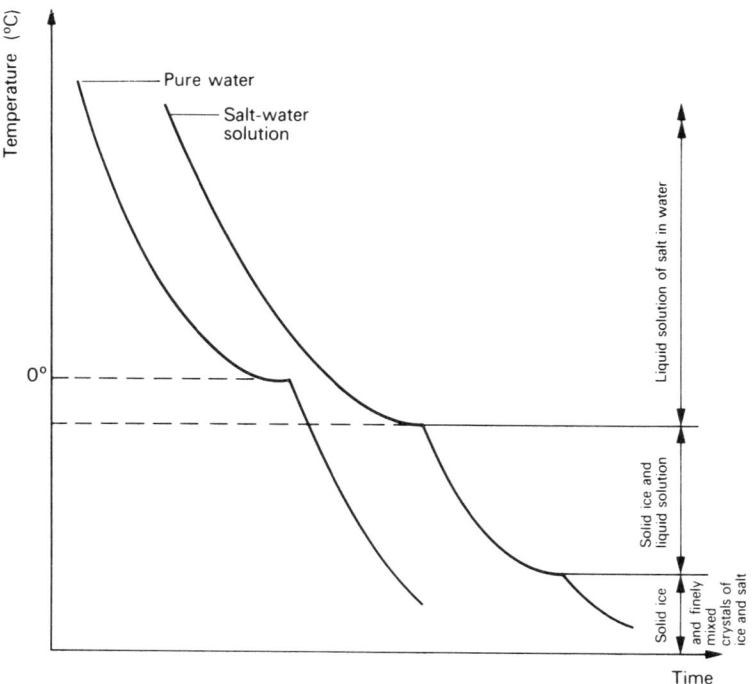

Fig. 2.6 Cooling curve for a salt-water solution

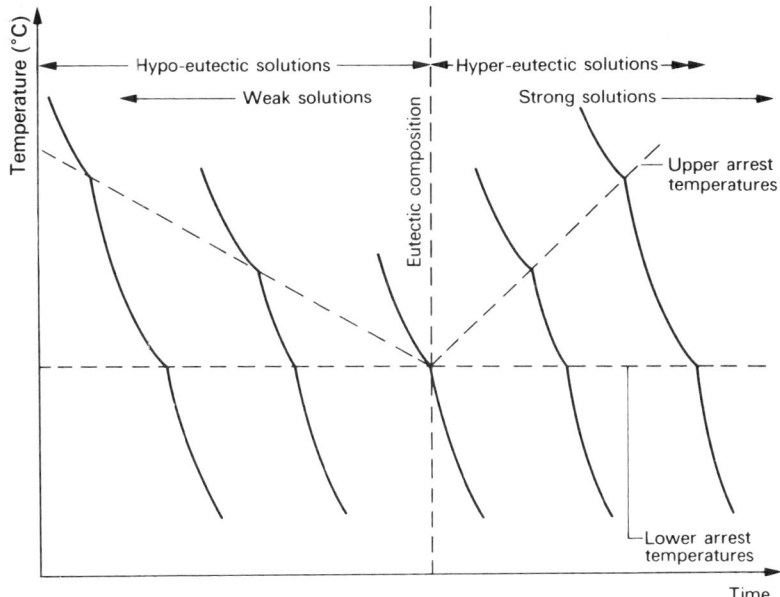

Fig. 2.7 'Family' of cooling curves

If the experiment is repeated several times using stronger and weaker salt-water solutions it will be seen that the upper arrest points vary as shown in Fig. 2.7, whilst the lower arrest points remain constant. The family of cooling curves so produced show some interesting trends. Reference to Fig. 2.7 shows that:

1. The temperature of the lower arrest point remains constant.
2. The temperature of the upper arrest point falls as the concentration of the solution increases until a point is reached where the temperatures of the upper and lower arrest points coincide.
3. The ratio of solid to liquid at the point where the temperatures coincide is referred to as a eutectic. Solutions with a lower concentration of solid to liquid are referred to as hypo-eutectic solutions. Solutions with a higher concentration of solid to liquid are referred to as hyper-eutectic solutions.
4. When the concentration of the solution increases beyond that of the eutectic composition the temperature of the upper arrest point rises once more.

Since water separates out as ice crystals between the arrest points of hypo-eutectic solutions, and since salt separates out between the arrest points of hyper-eutectic solutions, the remaining solution is always of a constant concentration. This concentration is the same as for the eutectic solution. The logic for this is apparent if reference is again made to Fig. 2.7. The fact that excess water or salt is rejected from the solution so that a eutectic 'balance' is always ultimately achieved, results in the diagram formed from the cooling curves (Fig. 2.7) being referred to as a thermal equilibrium diagram.

2.6 Alloy types

Alloys containing two components are referred to as *binary alloys*. Even when more than two components are present, a lot of useful information can be obtained from a study of the binary diagram of the two principal components present. The constituent components of most commercially available binary alloys are soluble in each other in the liquid (molten) state and, in general, do not form intermetallic compounds. (The exceptions are some bearing alloys.) However, upon cooling into the solid state, binary alloys can be classified into three main types.

1. *Simple eutectic type.* The two components are soluble in the liquid state, but completely *insoluble* in each other in the *solid state*.
2. *Solid solution type.* The two components are completely *soluble* in each other both in the liquid state and in the *solid state*.
3. *Combination type.* The two components are completely soluble in each other in the liquid state, but are only *partially soluble* in each other in the *solid state*. Thus this type combines some of the characteristics of both 1 and 2 above, hence the name 'combination type' thermal equilibrium diagram.

These three types of binary alloy and their thermal equilibrium diagrams will now be considered in greater detail.

2.7 Thermal equilibrium diagrams (eutectic type)

Figure 2.8 shows a eutectic-type thermal equilibrium diagram and it will be seen that it is identical with the diagram produced for a common salt solution (Fig. 2.7). That is, complete solubility of salt in water in the liquid state and complete insolubility (crystals of ice and separate crystals of salt) in the solid state. In the general case of Fig. 2.8 the two components present are referred to as metal A and metal B. In the solid state both components retain their individual identities as crystals of A and crystals of B.

Reference to Fig. 2.8 shows that the line joining the points where solidification begins is referred to as the *liquidus*. The line joining the points where solidification is complete is referred to as the *solidus*.

This type of equilibrium diagram gets its name from the fact that at one particular composition (E), the temperature at which solidification starts to occur is a minimum for the alloying components present. With this composition the liquidus and the solidus coincide at the same temperature, thus the liquid changes into a solid with both A crystals and B crystals forming instantaneously at the same temperature. This point on the diagram is called the *eutectic*; the temperature at which it occurs is the eutectic temperature, and the composition is the *eutectic composition* (see Fig. 2.10).

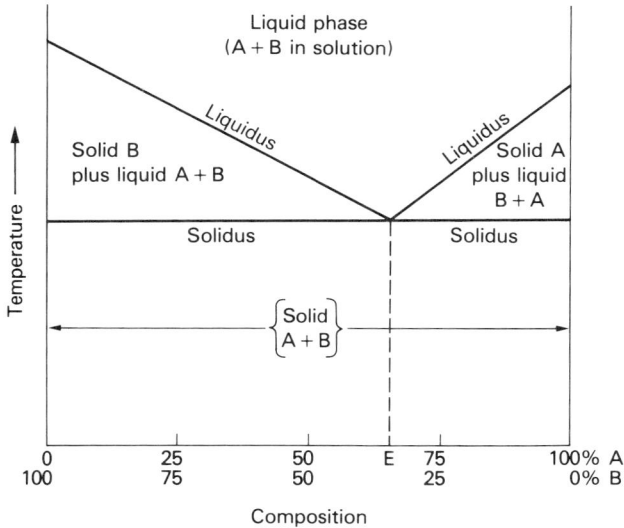

Fig. 2.8 Thermal equilibrium diagram (eutectic type)

Figure 2.8 also shows that when both alloying components are liquid (molten), this region of the diagram is referred to as the liquid *phase*. The term 'phase', when related to a thermal equilibrium diagram, is defined as a region on that diagram which has the same chemical composition or structure throughout. Thus above the liquidus the A and B components form a homogeneous solution and the definition is applicable. It is not applicable between the liquidus and solidus, neither is it applicable below the solidus in this diagram.

In practice, few metal alloys form simple eutectic type thermal equilibrium diagrams. Exceptions to this are the *cadmium-bismuth* alloys. The thermal equilibrium diagram for cadmium-bismuth alloys is shown in Fig. 2.9. It can be seen that the eutectic composition occurs when the alloy consists of 40 per cent cadmium and 60 per cent bismuth. Solidification occurs at just over 140°C with both metals crystallising out of solution simultaneously. The eutectic structure is usually *lamellar* in form, as shown in Fig. 2.10. In this instance there are alternate layers of cadmium and bismuth.

Consider the cooling of an alloy consisting of 80 per cent cadmium and 20 per cent bismuth (hyper-eutectic).

1. Above the liquidus there is a liquid solution of molten bismuth and molten cadmium.

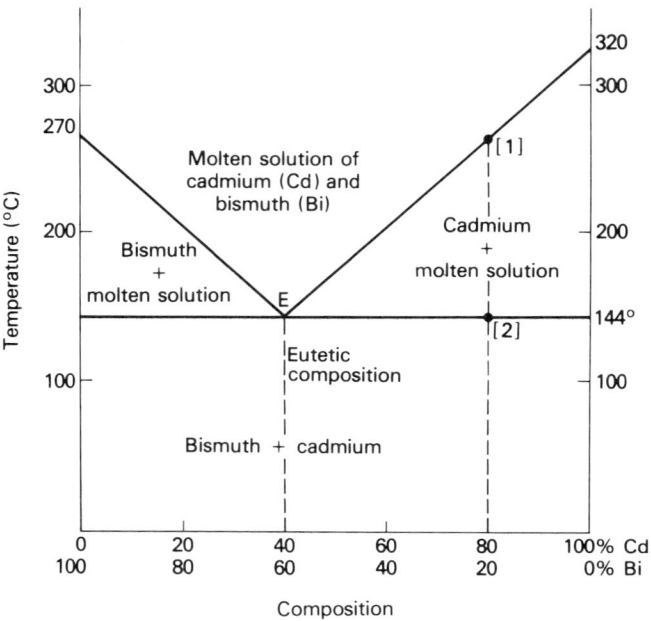

Fig. 2.9 Cadmium-bismuth thermal equilibrium diagram

2. As the solution cools to the liquidus temperature for the alloy under consideration crystals of pure cadmium start to precipitate out ([1] Fig. 2.9). This increases the concentration of bismuth and reduces the concentration of cadmium present in the remaining solution. Thus the solidification temperature is reduced to that appropriate for this new ratio of cadmium and bismuth, and further crystals of pure cadmium precipitate out. This again reduces the percentage of cadmium present in the remaining solution and the solidification temperature is further reduced with more pure cadmium crystals being precipitated out. This process repeats itself until the eutectic composition is reached ([2] Fig. 2.9).
3. At the eutectic composition, crystals of cadmium and bismuth precipitate out simultaneously to form lamellar eutectic crystals of the two metals as shown in Fig. 2.10. Thus the final composition of the solid alloy will consist of crystals of pure cadmium in a matrix of crystals of eutectic composition.

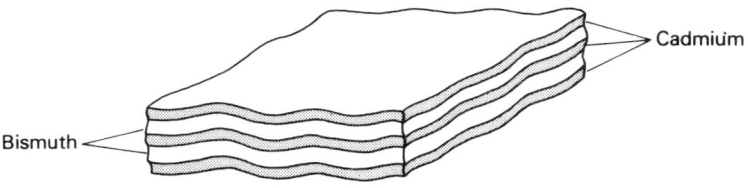

Fig. 2.10 Lamellar structure of eutectic composition

Similarly for an alloy of 80 per cent bismuth and 20 per cent cadmium (hypo-eutectic), the amount of cadmium present in solution compared with the amount of bismuth present in solution will gradually increase as pure crystals of bismuth precipitate out until the eutectoid composition is reached. Thus in this instance the composition of the solid alloy will consist of crystals of pure bismuth in a matrix of crystals of eutectic composition.

For an alloy of 60 per cent bismuth and 40 per cent cadmium only crystals of eutectic composition will be present. These solid alloy compositions are shown in Fig. 2.11.

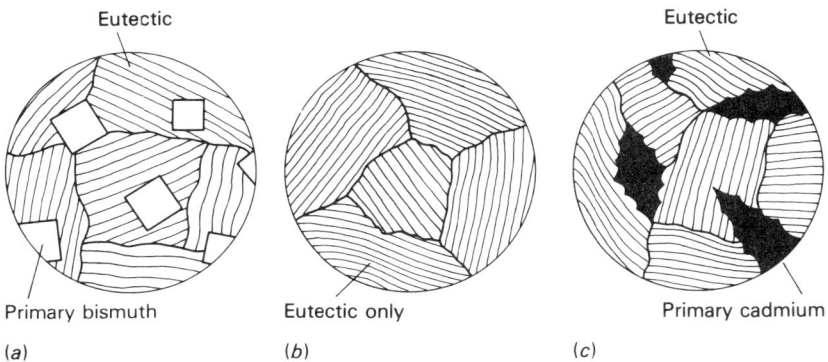

Fig. 2.11 Solid composition of cadmium-bismuth alloys (*a*) 20% Cd 80% Bi (*b*) 40% Cd 60% Bi (*c*) 80% Cd 20% Bi

2.8 Thermal equilibrium diagram (solid solution type)

It has already been stated that copper and nickel are not only mutually soluble in the liquid (molten) state, but are also mutually soluble in the solid state. They form a substitutional solid solution. The thermal equilibrium diagram for copper–nickel alloys is shown in Fig. 2.12. Again, the line marked liquidus joins all the points where solidification commences, whilst the line marked solidus joins all the points where solidification is complete. This time there is no eutectic composition.

Thus for 100 per cent copper, 0 per cent nickel (pure copper) there is a single solidification temperature of 1084 °C. This is to be expected since for a pure metal (in fact any pure substance) the transition from liquid to solid takes place at a constant temperature. For an alloy of 80 per cent copper and 20 per cent nickel Fig. 2.12 shows that solidification starts at 1190 °C and is complete at 1135 °C. Between the solidus and liquidus is a solution of molten copper and nickel together with crystals of a solid solution of copper and nickel. For an alloy of 80 per cent nickel and 20 per cent copper Fig. 2.12 shows that solidification starts at 1410 °C and is complete by 1380 °C. Finally, Fig. 2.12 shows that 100 per cent nickel, 0 per cent copper (pure nickel) solidifies at the single temperature of 1455 °C. Below the solidus the alloy consists entirely of crystals of copper and nickel in solid solution. Hence in this diagram it is correct to refer to the liquid phase and the solid phase.

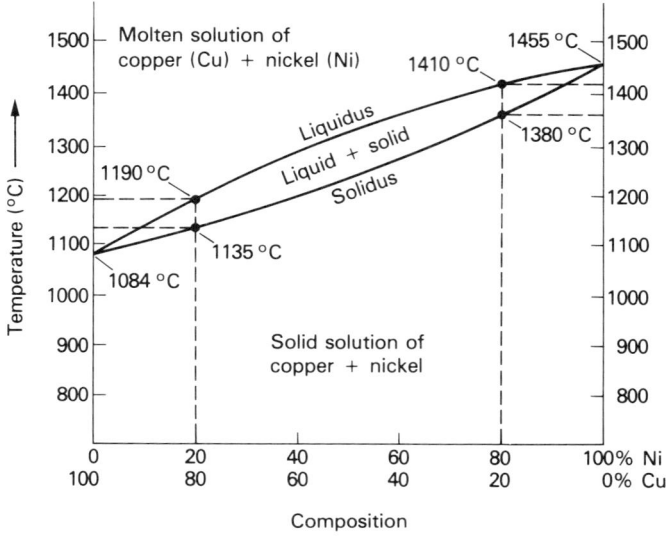

Fig. 2.12 Copper-nickel thermal equilibrium diagram

2.9 Thermal equilibrium diagram (combination type)

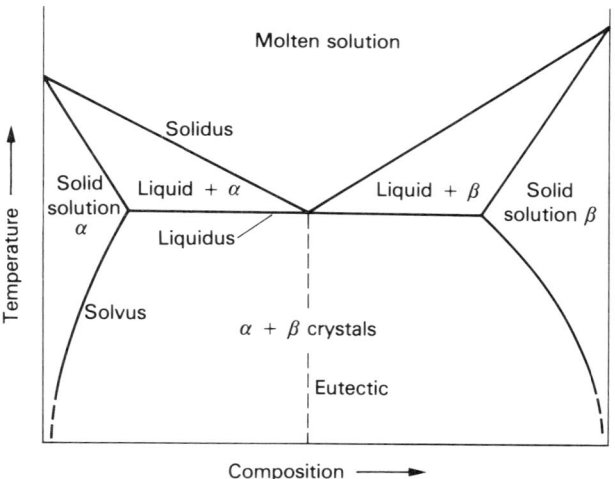

Fig. 2.13 Combination type thermal equilibrium diagram

Many metals and non-metals are neither completely soluble in each other in the solid state, nor are they completely insoluble. They form a thermal equilibrium diagram of the type shown in Fig. 2.13. In this system there are two solid solutions labelled α and β. Tin-lead alloys (soft solders) are a typical example of this type of thermal equilibrium diagram as shown in Fig. 2.14.

The use of the Greek letters α, β, etc. in thermal equilibrium diagrams may be defined, in general, as follows:

1. Solid solution of one component A in an excess of another component B such that A is the solute and B is the solvent.
2. Solid solution of the component B in an excess of component A so that now B becomes the solute and A becomes the solvent.

α Phase. This is a solid solution of 19.2 per cent tin in 80.8 per cent lead at the eutectic temperature.

β Phase. This is a solid solution of 2.6 per cent lead in 97.4 per cent tin at the eutectic temperature.

1. Above the liquidus ABC there is a homogeneous solution of molten tin and lead.
2. For hypo-eutectic alloys, the solidus is the line ADB. Between the liquidus and the solidus the hypo-eutectic alloys will consist of the liquid solution of tin and lead plus crystals of the solid solution of α composition.
3. Below the eutectic temperature, the line separating the α phase from the α + β phase is called the *solvus* (see Fig. 2.13).

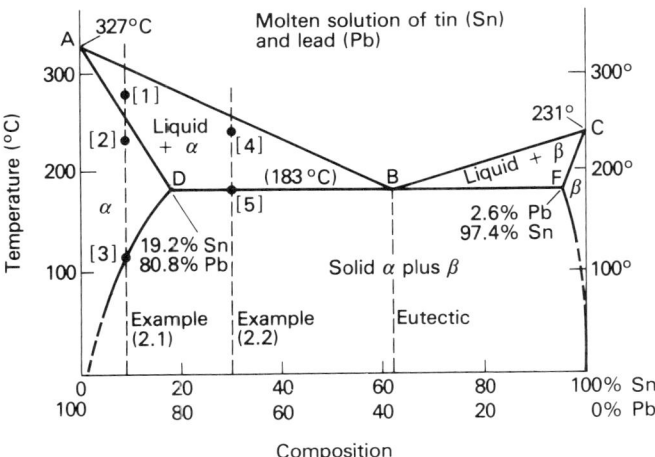

Fig. 2.14 Tin-lead equilibrium diagram

4. For hyper-eutectic alloys, the solidus is the line BFC. Between the liquidus and the solidus the hyper-eutectic alloys will consist of the liquid solution of tin and lead plus crystals of the solid solution of β composition.
5. Below the eutectic temperature, the line separating the β phase from the $\alpha + \beta$ phase is also called the solvus.

Example 2.1

Consider an alloy of composition 10 per cent tin, 90 per cent lead. Upon cooling from the molten state, where both metals are completely soluble in each other, to a temperature below the liquidus ([1] Fig. 2.14) then crystals of the α phase solid solution start to grow. As in the previous diagrams solidification is complete when the solidus is reached and the solid alloy will consist of crystals of the α phase in solid solution ([2] Fig. 2.14). The composition of this solid solution will be 19.2 per cent tin in 80.8 per cent lead, as previously stated. However, as the temperature of the alloy continues to fall it will eventually meet the solvus ([3] Fig. 2.14). At this point the solid solution will be *saturated* with tin. Further cooling to room (ambient) temperature will result in the tin precipitating out to form the other solid solution possible in this system, the β phase. Thus the final composition of this alloy will consist of tin-rich crystals of the β phase dispersed through a matrix of crystals of low tin content α phase.

Example 2.2

Consider an alloy of composition 30 per cent tin and 70 per cent lead. Upon cooling from the molten state, where both metals are completely soluble in each other, to below the liquidus ([4] Fig. 2.14) then crystals of α phase solid solution start to grow. This increases the concentration of tin and reduces the concentration of lead in the remaining molten solution. The solidification temperature is reduced to that appropriate for this new ratio, and the process repeats itself with more and more α phase solid solution being precipitated out until the eutectic composition is reached ([5] Fig. 2.14).

At this point crystals of both α and β phase solid solutions are precipitated out simultaneously to form lamellar eutectic crystals. (See Fig. 2.10). Thus the final composition of the solid alloy will consist of crystals of α phase solid solutions in a matrix of crystals of eutectic composition.

These are important examples to the engineer as they explain the behaviour of the various types of soft solder in popular use. Thus the popular 60 per cent tin, 40 per cent lead alloy known as 'tinman's' solder has the lowest melting and solidification temperature since it is approximately the eutectic alloy. This also accounts for its instant setting with no 'pasty' range. Its relatively high tin content and low electrical resistance also accounts for its widespread use for soldered joints in the electronics industry. On the other hand, the plumber requires a solder with a long 'pasty' range which will set slowly and enable a wiped joint to be made. A typical plumber's solder would consist of 80 per cent lead and 20 per cent tin so that there is a maximum temperature range between the liquidus and the solidus. At the same time the liquidus temperature is safely below the melting point of pure lead so that there is less danger of melting the lead pipe or component being joined.

There are many other examples of binary alloys which could be quoted, but the three examples considered cover the three most common types of thermal equilibrium diagram.

2.10 Coring

So far, cooling from above the liquidus has been assumed to be so slow that equilibrium is achieved as each change occurs. The cooling of a copper-nickel alloy under equilibrium conditions will now be considered in greater detail and the mechanism of solidification will then be compared with the same alloy cooled under production conditions.

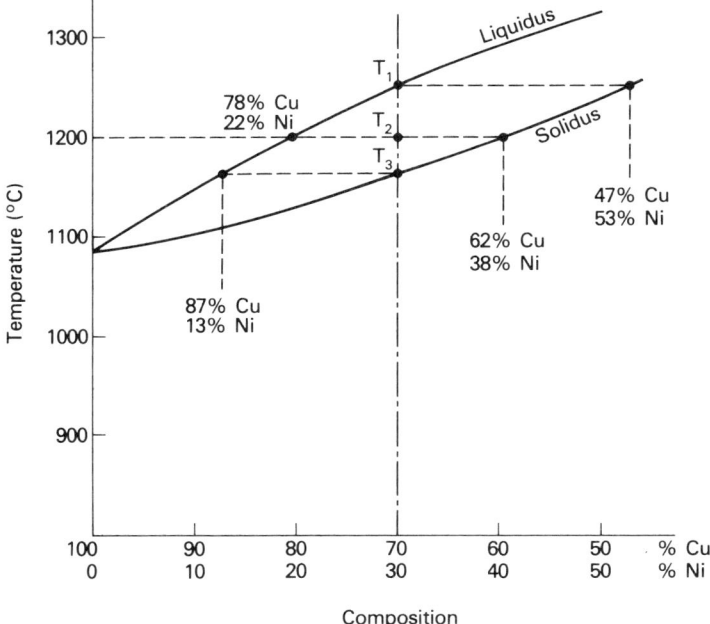

Fig. 2.15 Copper-rich Cu-Ni alloys

Figure 2.15 shows part of the copper–nickel thermal equilibrium diagram enlarged for clarity. It is convenient to consider an alloy of 70 per cent copper and 30 per cent nickel since solidification will conveniently centre on 1200 °C. When the molten alloy cools to the liquidus small dendrites of copper–nickel solid solution form. If a line is drawn from T_1 on the liquidus parallel to the composition axis until it cuts the solidus it is apparent that the composition of the solid solution will be 41 per cent copper and 53 per cent nickel. Since the overall composition of molten alloy is still 70 per cent copper and 30 per cent nickel, the fact that the dendrites have 53 per cent nickel means that the remaining molten solution will have less than 30 per cent nickel.

As the alloy cools down to 1200 °C, the dendrites grow in size. A line drawn through T_2 parallel to the composition axis until it cuts the solidus indicates that the composition of the solid solution for this temperature will be 62 per cent copper and 38 per cent nickel. Thus between T_1 and T_2 the percentage of copper present in the dendrite has increased, whilst the percentage of nickel present in the dendrite has fallen. Since the line through T_2 cuts the liquidus at 78 per cent copper and 22 per cent nickel, this is the composition of the remaining molten alloy.

Solidification is complete at T_3, with the composition of the solid solution 70 per cent copper and 30 per cent nickel. The line from T_3 to the liquidus indicates that the last drop of molten alloy will have a composition of 87 per cent copper and 13 per cent nickel. Thus it is apparent that the crystal will have a nickel-rich core and a copper-rich case unless something can restore the balance.

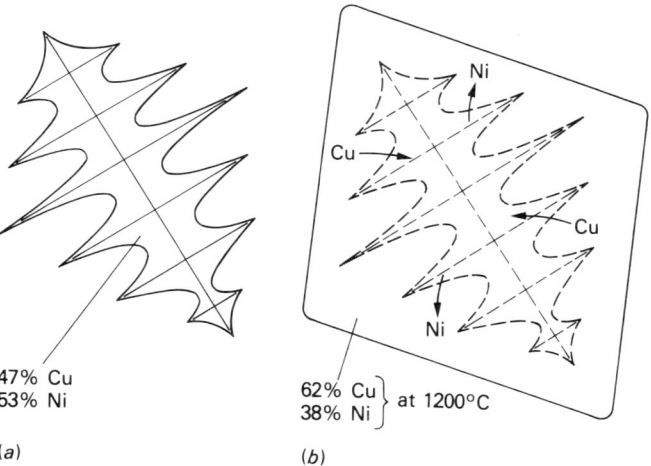

Fig. 2.16 Diffusion during crystal growth (*a*) Dendritic nucleus at liquidus temperature (*b*) Diffusion of copper and nickel as crystal commences to grow

If the entire process is slow enough so that *equilibrium* within the crystal is maintained from the start, then *diffusion* will occur with copper atoms migrating into the core of the crystal and nickel atoms migrating outwards into the case of the crystal. By the time cooling is complete, the composition should be uniform throughout with 70 per cent copper and 30 per cent nickel as shown in Fig. 2.16.

In thermal equilibrium diagrams it is always assumed that cooling will be slow enough for equilibrium to be maintained.

Under production conditions where cooling is more rapid than the ideal, for example the production of a copper–nickel casting, there is insufficient time for diffusion to become complete and the nickel-rich core will be apparent when an etched specimen is examined under the microscope. The core of the crystal will have the outline appearance of the initial dentrite from which the crystal has grown. The result of this more rapid cooling is called 'coring', and since coring leads to lack of uniformity in the structure of the metal, this adversely affects its mechanical properties.

Heat treatment of the solidified casting can eliminate coring. The process consists of heating the casting to just below the solidus for the alloy composition concerned, and holding it there until diffusion is complete. Once diffusion is complete the rate of cooling is irrelevant, except that over-fast cooling will create stresses in the metal.

2.11 Precipitation

Reference back to section 2.2 shows that the solubility of copper sulphate varies with temperature. In hot water at 80 °C it is possible to dissolve 50 g of copper sulphate in 100 g of water. On cooling to 40 °C the solubility falls to 30 g of copper sulphate per 100 g of water. The solution is now saturated and 20 g of copper sulphate is *precipitated* out of solution.

A similar effect can occur in solid solutions and advantage is taken of this in the heat treatment of some non-ferrous alloys. Notably those containing aluminium and copper, and those containing magnesium and aluminium.

Figure 2.17 shows part of the aluminium–copper thermal equilibrium diagram. If an alloy containing 4 per cent copper is cooled from the molten condition to 500°C (T_1) the alloy consists of crystals of solid solution. The solid solution does not become saturated until the solvus is reached at temperature T_2. If cooling continues slowly (equilibrium conditions) the crystals of solid solution will remain saturated but some precipitation will occur. At room temperature (T_3) the structure of the alloy will consist of crystals of solid solution containing a coarse precipitate of $CuAl_2$. In this condition the alloy will be soft, but relatively weak.

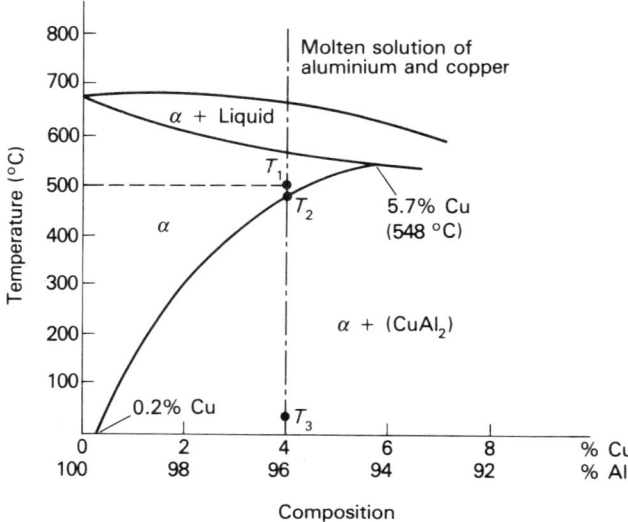

Fig. 2.17 The solution treatment of aluminium-copper alloys

If, on the other hand, the alloy is quenched from temperature T_1 so that it cools quickly there is not time for the equilibrium conditions to be achieved. This is because it requires a grouping together of atoms by diffusion for precipitation to occur and this is a slow process. The result of such rapid cooling is to prevent precipitation so that the crystals consist of a *supersaturated* solid solution of α phase alloy at room temperature. This is referred to as *solution treatment* and is the way in which such alloys may be 'annealed' or softened ready for cold working. In this condition the alloy is slightly harder and tougher than when cooled under equilibrium conditions.

The supersaturated solid solution can be retained at room temperature, but it is not a stable condition and precipitation can occur with the elapse of time. This occurs in the aluminium–copper alloy 'duralumin' and this alloy has to be kept under refrigerated conditions to delay the onset of precipitation. The precipitate tends to be fine particles evenly distributed throughout the crystals of solid solution. These fine particles give a harder and stronger alloy than when precipitation occurs under equilibrium conditions.

When precipitation occurs from the supersaturated condition with the elapse of time it is referred to as *Natural ageing*. Where precipitation is brought about by reheating the alloy in a furnace at about 165 °C for several hours, the process is referred to as *Artificial ageing*. Both processes are referred to as *precipitation age hardening*.

Problems

Section A
1. Describe the difference between a pure metal and an 'alloy'.
2. Explain what is meant by the terms: (i) solvent; (ii) solute; (iii) solution; (iv) solid solution.
3. With the aid of sketches show the difference between a *substitutional* solid solution; and an *interstitial* solid solution.
4. What is meant by an 'inter-metallic' compound and how are its properties likely to vary from those of a solid solution?
5. Sketch typical examples of the following binary, thermal equilibrium diagrams: (i) simple eutectic type; (ii) Solid solution type; (iii) Combination type.

Section B
6. (a) Sketch typical cooling curves for pure water and for a sodium chloride solution from 80 °C down to −10 °C and comment upon their relative shapes.
   (b) Show how a 'thermal equilibrium' may be derived from the cooling curves for a sodium chloride solution and clearly indicate: (i) the liquidus; (ii) the solidus; (iii) the eutectic point.
7. Draw the thermal equilibrium diagram for cadmium–bismuth alloys.
   (a) State the eutectic composition and describe the structure of a eutectic alloy in the solid state.
   (b) Explain in detail the cooling of an alloy of composition 90 per cent cadmium and 10 per cent bismuth, and describe its structure in the solid state.
8. Draw the thermal equilibrium diagram for copper–nickel alloys and explain in detail the cooling of an alloy of composition 40 per cent nickel and 60 per cent copper. Explain why this series of alloys do not show a eutectic composition.
9. Draw the thermal equilibrium diagrams for tin–lead alloys.
   (a) Indicate on the diagrams: (i) the liquidus; (ii) the solidus; (iii) the solvus.
   (b) state the eutectic composition and describe the structure of the eutectic alloy in the solid state.
   (c) Explain in detail the cooling of an alloy of 40 per cent tin and 60 per cent lead and describe its structure in the solid state.
   (d) Explain briefly why 'plumbers' solder has a high lead content, whilst a 'tinman's' solder has a eutectic composition.
10. With reference to the copper–nickel thermal equilibrium diagram, explain what is meant by 'coring' and 'diffusion'.
11. With reference to the aluminium–copper thermal equilibrium diagram, explain what is meant by 'solution treatment' and 'precipitation hardening'.

Chapter 3

Plain carbon steels

3.1 Ferrous metals

Ferrous metals and alloys are based upon the metallic element *iron*. The name ferrous comes from the Latin name for iron which is *ferrum*. Iron is a soft, grey metal and it is rarely found in the pure state outside the laboratory. The engineer usually finds it associated with the non-metal *carbon*, with which it forms solid solutions and also the compound iron carbide. To extract the metal iron from its ore a chemical reaction known as a *reduction* has to be used. The reducing agent used in this reaction has to be a carbon-rich material, such as coke in the traditional blast furnace or natural gas in the more modern processes.

Since all the ferrous materials used by engineers contain iron in association with carbon, it could be argued that all such materials are ferrous alloys. However, this term is reserved for those ferrous materials containing additional metallic alloying elements in sufficient quantities substantially to modify the properties of the material. For example: nickel–chrome alloy steel, or chrome–vanadium alloy steel, etc. Those 'alloys' containing only *carbon* as the main alloying element are referred to as wrought iron, plain carbon steels and cast irons. Table 3.1 shows the relationship between the amount of carbon present and the ferrous metal produced. It also gives typical applications of those metals. The plain carbon steels will be considered in this chapter and the cast irons in Chapter 6. The alloy steels will be considered in *Materials technology: level 3*.

Table 3.1 Ferrous metals

Name	Group	Carbon content %	Some uses
Wrought iron	Wrought iron	Less than 0.05	Chain for lifting tackle, crane hooks, architectural ironwork.
Dead mild steel	Plain carbon steel	0.1 to 0.15	Sheet for pressing out such shapes as motor car body panels. Thin wire, rod, and drawn tubes.
Mild steel	Plain carbon steel	0.15 to 0.3	General purpose workshop bars, boiler plate, girders
Medium carbon steel	Plain carbon steel	0.3 to 0.5 0.5 to 0.8	Crankshaft forgings, axles Leaf springs, cold chisels
High carbon steel	Plain carbon steel	0.8 to 1.0 1.0 to 1.2 1.2 to 1.4	Coil springs, wood chisels Files, drills, taps and dies Fine-edge tools (knives, etc.)
Grey cast iron	Cast iron	3.2 to 3.5	Machine castings

3.2 The iron–carbon system

Figure 3.1 (*a*) shows the iron–carbon thermal equilibrium diagram. Strictly, it should be called the iron–carbide diagram, but conventionally it is called the iron–carbon diagram. Comparison with the diagrams described in Chapter 2 shows it to be of the type where two substances are completely soluble in each other in the liquid (molten) state, but only partially soluble in the solid state. The diagram is modified from those considered in Chapter 2 by the structural changes which take place at 910 °C and 1400 °C. These are due to the fact that iron is *allotropic*, that is, it can exist in more than one form.

(*a*) Below 910 °C the iron forms body-centred cubic crystals.
(*b*) From 910 °C to 1400 °C it forms face-centre cubic crystals.
(*c*) Above 1400 °C it reverts to body-centred cubic crystals.

These changes are accompanied by changes in *volume* as the atoms in the crystal lattice rearrange themselves. For example, when iron is heated, it expands uniformly with temperature until 910 °C, whereupon it contracts slightly as the atoms rearrange themselves into a more compact lattice, after which it continues to expand again. (See Fig. 3.1 (*b*).

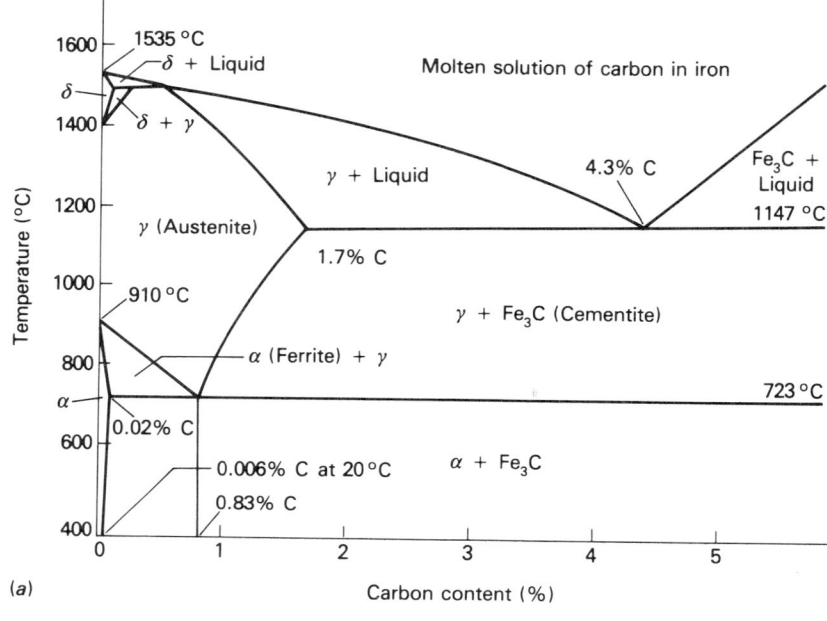

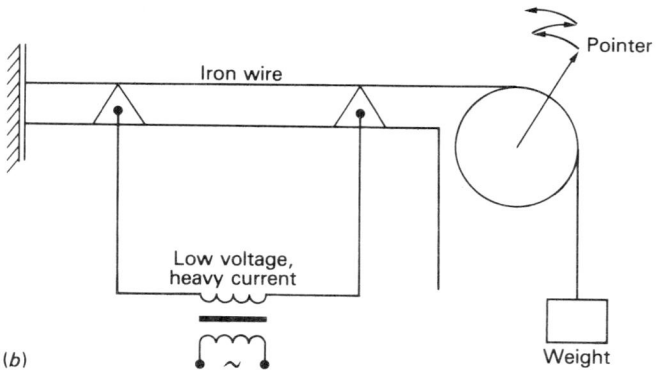

Fig. 3.1 The phase changes of iron-carbon alloys (*a*) Iron-carbon thermal equilibrium diagram (*b*) Method of demonstrating changes in volume as crystal lattices, rearrange themselves

The structural changes are also accompanied by *latent heat* energy being taken in or given out. If an iron rod is cooled from above 910°C in a darkened room, it will be seen suddenly to glow again with increased brightness. The latent heat energy given out as the crystal lattice changes from face-centred to body-centred results in a momentary rise in temperature, and this causes the rod to glow more brightly.

This phenomenon is called recalescence.

The iron–carbon equilibrium diagram shown in Fig. 3.1 appears to be very complex compared with those considered in Chapter 2. Fortunately, in this chapter, we are only concerned with the 'steel section' of this diagram. This steel section of the iron–carbon diagram has been redrawn to larger scale in Fig. 3.2. It can be seen from Fig. 3.2 that there are only three important phases:

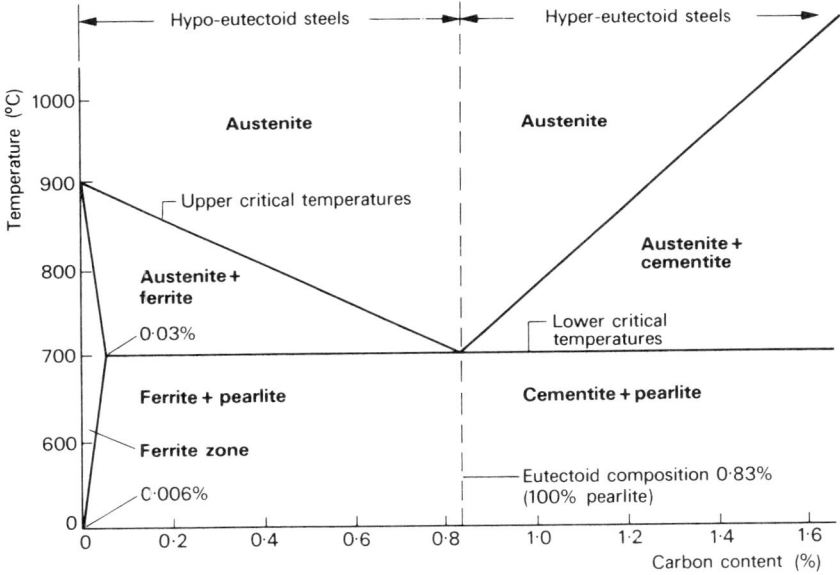

Fig. 3.2 Iron-carbon equilibrium diagram (steel section)

1. *Ferrite* (α phase). This is a weak solid solution of carbon in body-centred cubic crystals of iron. There is a maximum of 0.03 per cent carbon in solid solution at 723 °C, falling to 0.006 per cent carbon in solid solution at room temperature. Ferrite is very soft and ductile, and of relatively low strength.
2. *Austenite* (γ phase). This is a more concentrated solid solution than ferrite. It is formed when carbon dissolves in face-centred cubic crystals of iron. The maximum amount of carbon which can be held in solid solution with iron is approximately 1.7 per cent at 1150 °C, see Fig. 3.1. This is the upper limit of carbon which can be present in plain carbon steels. However, for practical purposes there is no advantage in increasing the carbon content beyond about 1.2 to 1.4 per cent.

3. *Cementite* (iron–carbide phase). An excess of carbon (C) combines with iron (Fe) to form carbide (Fe_3C). Each molecule of iron carbide contains three atoms of iron and one atom of carbon. It is very hard and brittle, and lacks ductility.

SUMMARY

δ – iron	B.C.C.	low solubility of carbon in iron 1400 °C
γ – iron	F.C.C.	higher solubility of carbon in iron (1.7 per cent) 910 °C
α – iron (ferrite)	B.C.C.	low solubility of carbon in iron (0.03 per cent at 910 °C falling to 0.006 per cent at 0 °C)

Cementite – an excess of carbon forming molecules of iron carbide (Fe_3C), 3 atoms of iron to 1 atom of carbon

Pearlite – a lamellar structure of ferrite and cementite with a eutectoid composition of 0.83 per cent carbon and 99.17 per cent iron

---

The steel section of the iron–carbide thermal equilibrium diagram is very similar in appearance to the eutectic type thermal equilibrium diagram described in Chapter 2. In the eutectic type diagram there was one composition at which both alloying elements crystallised out simultaneously at the same temperature to form a lamellar structure. However, in the steel section of the iron–carbon thermal equilibrium diagram such transformations occur in the solid and the point at which ferrite and cementite precipitate out simultaneously from austenite is called the *eutectoid* point. This point occurs at 0.83 per cent carbon and a temperature of 723 °C. Crystals are produced with lamellar structure of alternate layers of ferrite and cementite. This structure is referred to as *pearlite*. It is the toughest structure possible in a plain carbon steel. At 0.83 per cent carbon the steel consists entirely of pearlite. Steels with a carbon content below 0.83 per cent are called *hypo-eutectoid* steels, whereas steels with a carbon content above 0.83 per cent carbon are called *hyper-eutectoid* steels.

The transformations which occur during the cooling of a eutectoid composition (0.83 per cent carbon) steel are shown in Fig. 3.3 (*a*). The steel commences to solidify at the liquidus (T_1) and solidification is complete when the solidus is reached at temperature (T_2). The steel now consists entirely of γ phase crystals of austenite. At 723°C (T_3) the austenite suddenly changes simultaneously into pearlite as shown. It remains as pearlite at all temperatures below 723°C. Figure 3.3 (*b*) shows a microphotograph of lamellar pearlite and the individual layers are clearly shown.

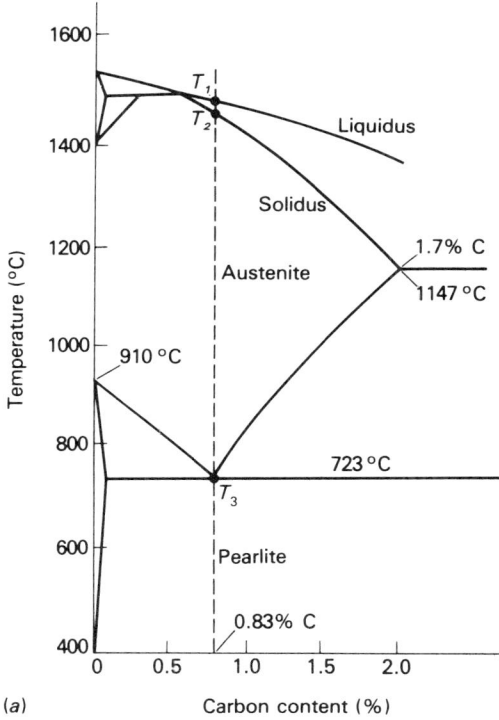

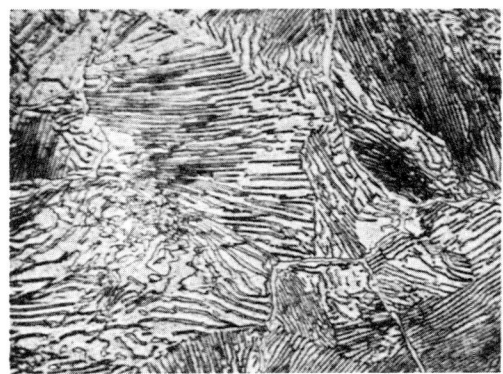

Fig. 3.3 Transformations for 0.83% carbon steel (a) Transformations at the eutectoid composition (b) Lamellar pearlite (× 600)

The transformations which occur during the cooling of a hypo-eutectoid steel are shown in Fig. 3.4 (a). In this example the steel contains 0.5 per cent carbon. Again the steel will be entirely austenitic once the solidus has been attained (T_2) and will consist entirely of crystals of γ phase austenite. However, on cooling slowly below (T_3) crystals of ferrite will start to grow in the austenite so that both α phase and γ phase crystals are present. Since α phase crystals of ferrite will contain rather less than 0.03 per cent carbon in solid solution, the carbon content of the remaining phase austenite will increase progressively as more and more ferrite is formed until at 723°C (the eutectoid temperature) the structure will contain 0.03 per cent carbon ferrite, and austenite containing 0.83 per cent carbon. Thus at T_4 the austenite will change suddenly into the eutectoid composition of pearlite. Thus the final composition of the steel below (T_4) will consist of crystals of ferrite and pearlite. Figure 3.4 (b) shows a typical microstructure for an annealed 0.5 per cent carbon steel.

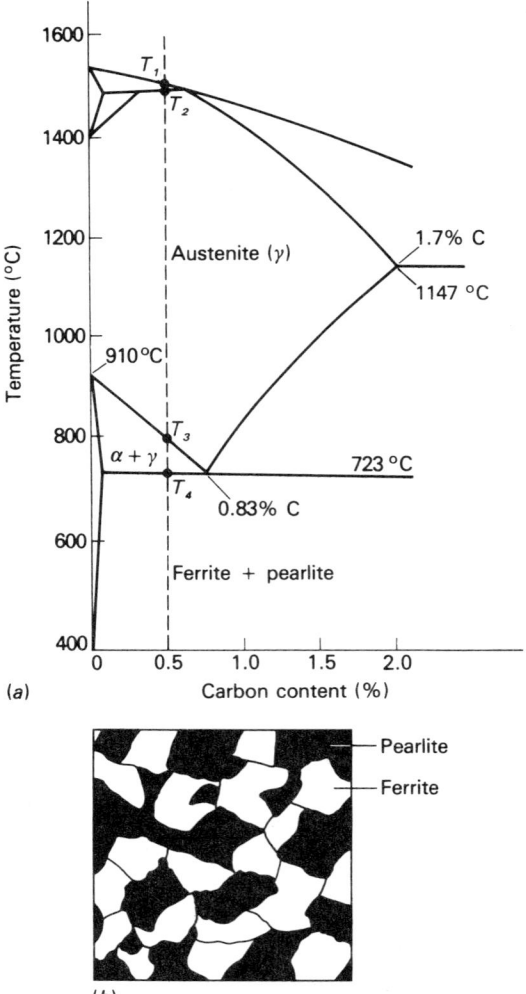

Fig. 3.4 Transformations for a 0.5% carbon steel (a) Transformations for hypoeutectoid steels (b) Typical microstructure for a 0.5% carbon steel. (Annealed)

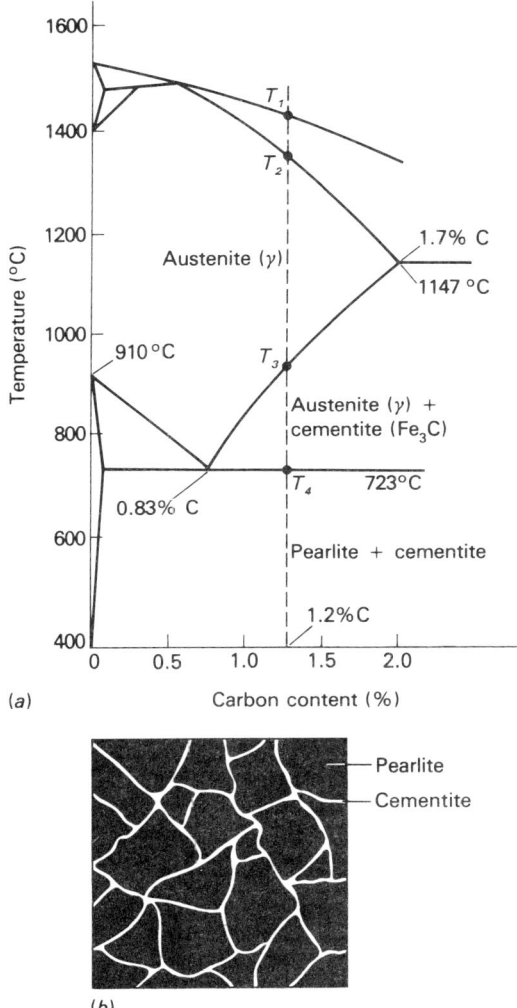

Fig. 3.5 Transformations for 1.5% carbon steel (*a*) Transformations for hypereutectoid steels (*b*) Typical microstructure for a 1.5% carbon steel. (Annealed)

The transformations which occur during the cooling of a hypereutectoid steel are shown in Fig. 3.5 (*a*). Again, the structure is entirely austenitic with crystals of the γ phase below temperature (T_2). Upon cooling below temperature (T_3) needles of primary cementite (iron–carbide) will start to precipitate out. Since cementite contains 1.7 per cent carbon, the remaining austenite will be less rich in iron. Eventually, at 723 °C, the carbon content of the austenite will have been reduced to 0.83 per cent and it will be transformed into crystals of pearlite with a eutectoid composition. Thus the final composition of the steel will consist of crystals of pearlite surrounded by bands of primary cementite at the crystal boundaries. Figure 3.5 (*b*) shows a typical microstructure for an annealed 1.5 per cent carbon steel.

3.3 Critical change points

The construction of thermal equilibrium diagrams from a family of cooling curves was explained in Chapter 2. The iron–carbon equilibrium diagram was constructed from just such a family of cooling curves by connecting the critical change points. These change points are often referred to, simplistically, as the upper critical temperature (U.C.T.) and the lower critical temperature (L.C.T.). The lower critical temperature is the eutectoid temperature of 723 °C, whilst the upper critical temperature depends upon the carbon content of the steel.

The critical change points, where changes in composition and structure occur, are also called *arrest* points since the time–temperature heating or cooling curve stops at these points as the latent heat energy associated with change is taken in during heating, or given out during cooling (see Fig. 3.6).

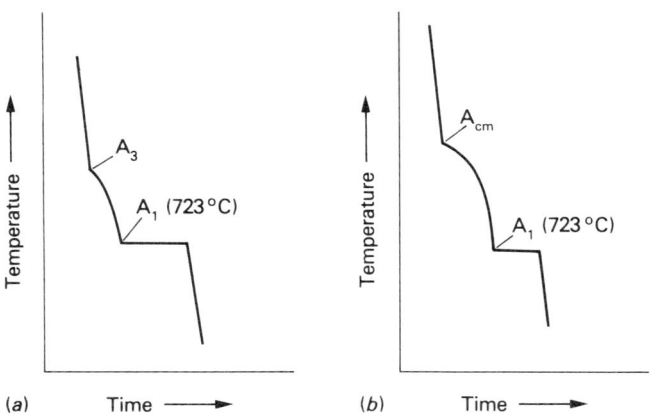

Fig. 3.6 Cooling curves for carbon steels (*a*) Cooling curve for a hypo-eutectoid steel (*b*) Cooling curve for a hyper-eutectoid steel

A_1 is the temperature at which the eutectoid transformations take place; that is the transformation of austenite into pearlite on cooling and vice versa on heating.

A_3 is the temperature above which hypo-eutectoid steels are wholly austenitic (γ phase).

A_{cm} is the temperature above which hyper-eutectoid steels are wholly austenitic (γ phase).

Due to what is known as thermal inertia the arrest points do not occur at exactly the same temperature on heating curves as they do on cooling curves. Therefore the arrest points require further identification to indicate whether they are heating or cooling points. This further notation makes use of the French word for heating which is *chauffage*. Thus the critical change points on a time–temperature heating curve are called: Ac_1, Ac_3, Ac_{cm}.

Similarly the French word for cooling, which is *refroidissement*, is used for time–temperature cooling curves. The critical points are then called: Ar_1, Ar_3, Ar_{cm}.

Their disposition on the thermal equilibrium diagram for plain carbon steels is shown in Fig. 3.7. Since the thermal equilibrium diagram is generally only used by engineers for determining heat treatment criteria (see Chapter 4) the cooling diagram based on *Ar* temperatures is the one usually quoted in engineering texts.

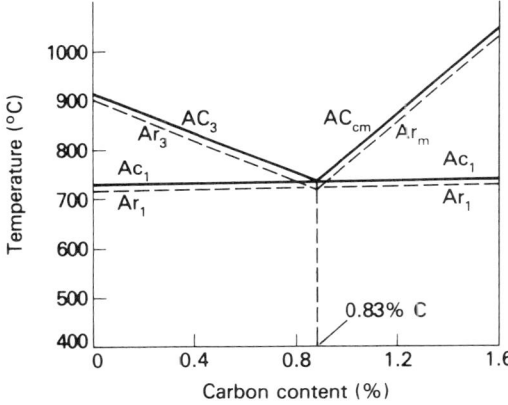

Fig. 3.7 Critical change points for carbon steels

There is also an A_2 temperature lying between A_1 and A_3, and this is the temperature at which the steel ceases to be magnetic. Since this does not affect the mechanical properties of the steel it is not normally included on the thermal equilibrium diagram, to avoid confusion. The A_2 temperature is often referred to as the *Curie point* after the French physicist who discovered it.

3.4 The effect of carbon on the properties of plain carbon steels

It has already been shown in section 3.2 that in a steel which has been cooled slowly enough to enable it to reach structural equilibrium, one of the following structures will be found.

1. With less than 0.03 per cent carbon only ferrite (α phase) will be present. This is commercially pure iron.
2. Between 0.03 per cent carbon and 0.83 per cent carbon the structure will contain ferrite and pearlite. The relative amounts of ferrite and pearlite present will vary according to the carbon content (see Fig. 3.8).
3. With exactly 0.83 per cent carbon (eutectoid composition) the structure will consist entirely of pearlite.
4. Between 0.83 per cent carbon and 1.7 per cent carbon the structure will consist of pearlite and cementite. The relative amounts of pearlite and cementite present will vary according to the carbon content (see Fig. 3.8).

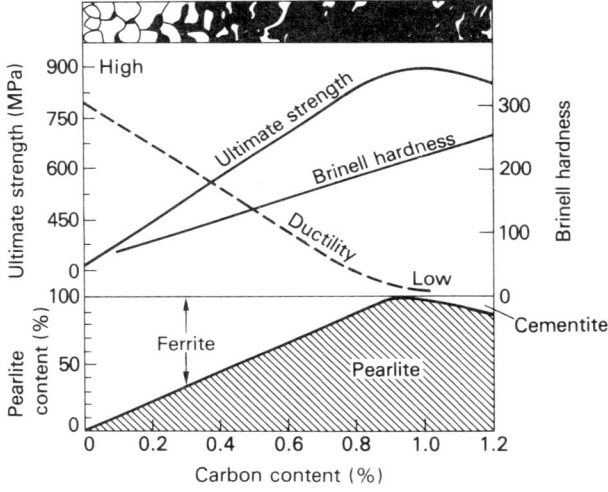

Fig. 3.8 Properties of plain carbon steels

Figure 3.8 shows the effect of the carbon content upon the properties of plain carbon steels. It has been shown that the upper limit of carbon content is 1.7 per cent. This is the maximum amount of carbon that can combine with iron at room temperature. In practice there is little advantage in raising the carbon content above 1.4 per cent.

It will be seen from Fig. 3.8 that low carbon steels are relatively soft, weak and ductile. This is because they consist mainly of crystals of ferrite, and ferrite is relatively weak, soft and ductile.

The increased amount of carbon in medium carbon steels promotes the formation of cementite. This results in an increased presence of pearlite, making such steels strong and tough but not so ductile as the low carbon steels.

When the carbon content reaches approximately 0.83 per cent the steel consists entirely of pearlite. This is the eutectoid composition, and it produces plain carbon steel of maximum toughness and strength.

Increasing the carbon content still further increases the amount of iron carbide or cementite in the steel. Since the maximum amount of combined cementite occurred at 0.83 per cent carbon, the formation of further cementite results in it appearing around the crystal boundaries. This reduces the ductility of the steel, but increases its hardness.

3.5 Wrought iron

Wrought irons are a group of ferrous metals with such a low carbon content that the iron/carbon compounds essential to a steel cannot be formed. In addition they contain fibres of slag trapped in the metal that promote corrosion resistance and prevent sudden fractures occurring. Table 3.2 lists the properties and some typical applications for wrought iron.

Table 3.2 Wrought iron

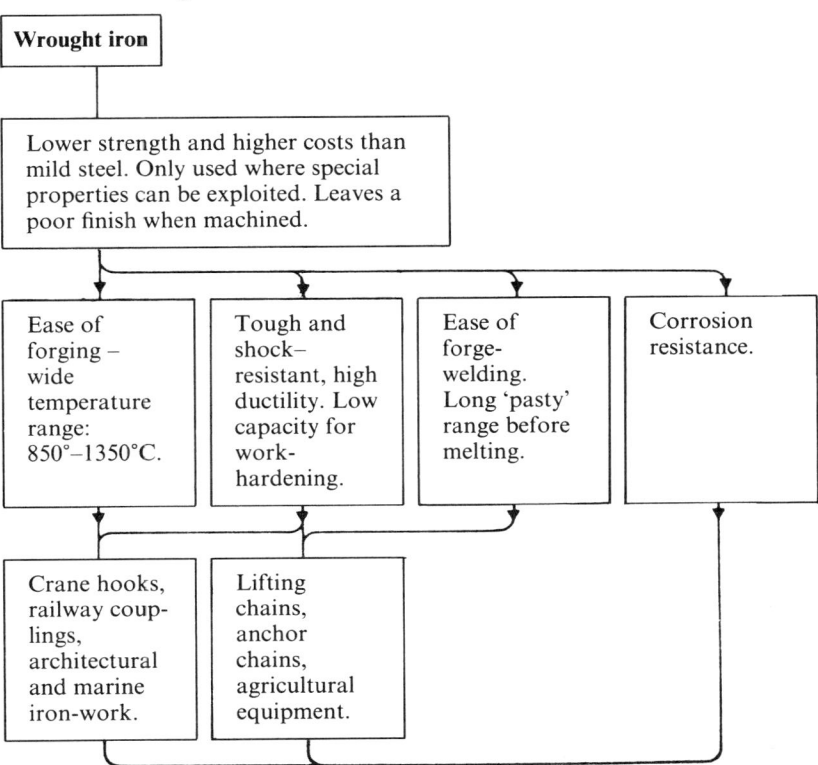

3.6 Plain carbon steels

These are ferrous materials containing between 0.1 per cent and 1.7 per cent carbon as the main alloying element. In addition to the carbon all plain carbon steels contain the following elements either by accident or by design.

Manganese: up to 1.0%
Silicon: up to 0.3%
Sulphur: up to 0.05%
Phosphorus: up to 0.05%

Manganese

Manganese is an essential constituent element since it ensures a sound ingot free from blow holes. Further, it combines with any sulphur present which would otherwise weaken the steel. In general, manganese raises the yield point, together with the strength and toughness of the steel. However, it increases the tendency of the steel to crack and distort when quench hardened and for this reason the content should be kept below 0.5 per cent in medium and high carbon steels.

Silicon

Silicon is an impurity carried over from the ore. It should be limited to 0.1 or 0.2 per cent in steels otherwise it can cause breakdown of the cementite which would result in weakness. Silicon has little direct effect upon the mechanical properties providing the amount present is limited to the percentage quoted above. It is often added to cast irons to prevent chill hardening (see section 2.9).

Sulphur

Sulphur is an impurity carried over from the coke used in the blast furnace. It tends to combine with the iron to form ferrous sulphide which greatly weakens the steel. Fortunately it has a greater affinity for manganese, and manganese sulphide does not weaken the steel. For this reason the amount of sulphur present should be kept below 0.05 per cent and there should always be at least five times as much manganese present as there is sulphur. Some free-cutting steels contain up to 0.2 per cent of sulphur to improve their machinability at the expense of strength for lightly stressed turned parts.

Phosphorus

Phosphorus is an impurity carried over from the ore. It forms compounds that make the steel brittle and, therefore, should be refined out as far as possible. It should not exceed 0.5 per cent.

Dead mild steel

The carbon content is deliberately left low so that the steel will have a high ductility. This enables it to be pressed into complicated shapes even while it is cold. It is slightly weaker than mild steel and is not usually machined since its softness would cause it to tear and leave a poor finish. It is used extensively for motor car body panels.

Mild steel

This is relatively soft and ductile, it can be forged and drawn in the hot or cold conditions, and it is easily machined using high-speed steel cutting tools (see Table 3.3).

Table 3.3 Mild steel

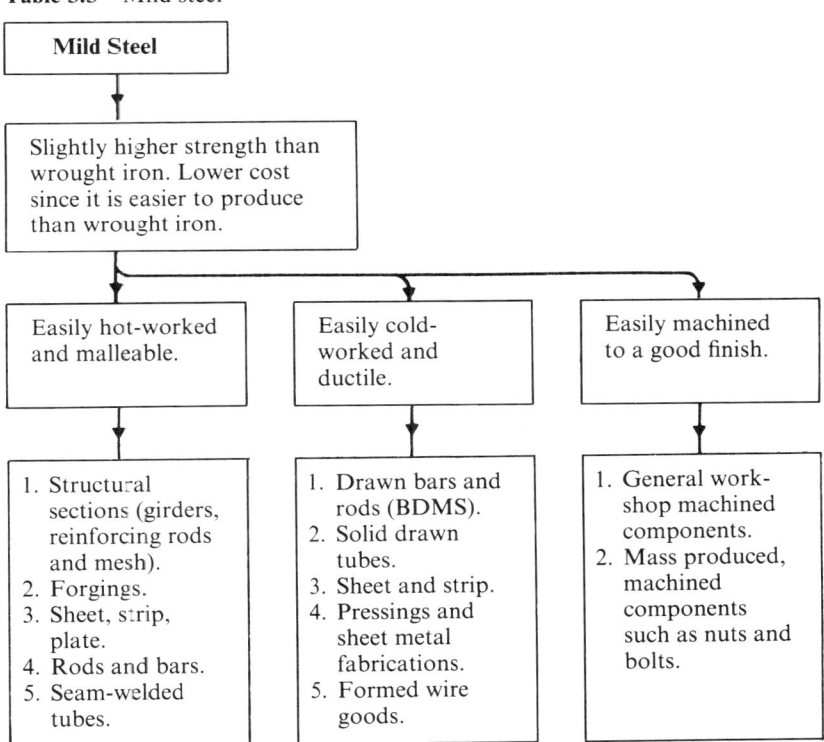

Medium carbon steel

This is harder, tougher and less ductile than mild steel, and cannot be bent or formed in the cold condition to any great extent without cracking. It hot forges well, but close temperature control is required to prevent:

1. 'burning' at high temperatures (over 1150°C), which leads to embrittlement;
2. cracking below 700°C, due to work hardening (see Table 3.4).

Table 3.4 Medium carbon steel

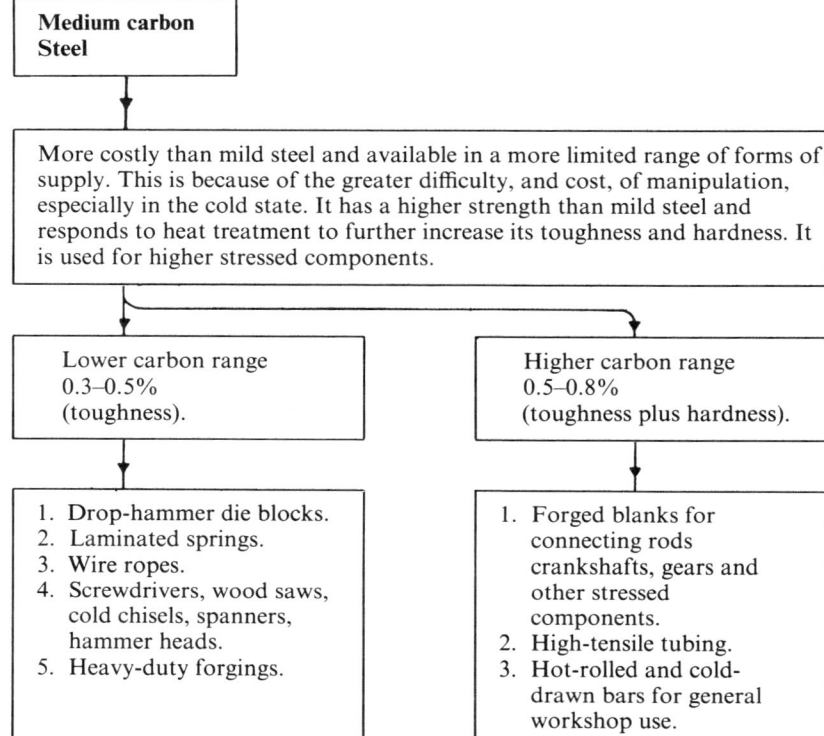

High carbon steel

This is harder, less ductile and slightly less tough than medium carbon steel. Cold forming is not recommended, but it hot forges well, providing the temperature is even more closely controlled, with an upper limit of 900 °C and a low limit of 700 °C (see Table 3.5).

Table 3.6 gives the composition and properties of some typical plain carbon steels, together with reference to their British Standard Specification. In comparing the properties of these steels it should be noted that some have had their properties modified by heat treatment. The state of the steel is, therefore, also stated in the table.

Table 3.5 High carbon steel

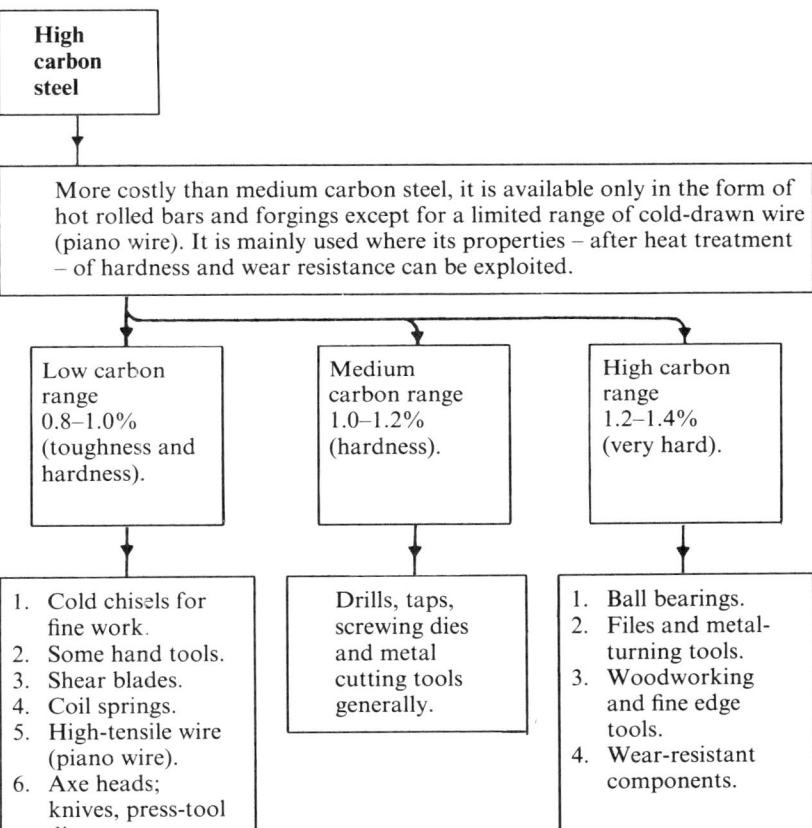

Table 3.6 Some plain carbon steels

Type of Steel	British Standard Specifications	Composition C%	Composition Mn%	Condition	Properties Y.P. (MPa)	U.T.S. (MPa)	Elong^n (%)	Impact (J)	Hardness (H_B)	Applications
'Dead' mild steel	BS 970.040A10	0.10	0.40	Process annealed after cold rolling	—	300	28	—	—	Car body panels produced by drawing and pressing.
Mild steel (structural)	BS 15 BS968	0.20 0.20	— 1.50	As rolled As rolled	240 350	450 525	25 20	— —	— —	General purpose: mild steel. Welding quality, high tensile mild steel for building construction, etc.
Casting steel	BS 1504/161B	0.30	—	Annealed after casting to refine grain.	265	500	18	20	150	General purpose, medium strength castings for machining.
Medium carbon steel	BS 970.080M40	0.40	0.80	Toughened by quenching from 850°C, temperature 550 600°C.	500	700	20	55	200	Axles, crankshafts, etc., under moderate stress.
	BS 970.070M55	0.55	0.70	Harden by quenching from 825°C. Temper at 600°C	550	750	14	—	—	Gears and machine parts subject to wear.

Table 3.6 continued Some plain carbon steels

Type of Steel	British Standard Specifications	Composition		Condition	Properties					Applications
		C%	Mn%		Y.P. (MPa)	U.T.S. (MPa)	Elong^n (%)	Impact (J)	Hardness (H_B)	
High Carbon Steels	—	0.70	0.35	Quench harden from 790/810 °C in water. Temper at 150 to 300 °C as appropriate.	—	—	—	—	780	Hand chisels, cold sets, screwdriver blades, blacksmith's tools, etc.
	BS4659: BW18	1.00	0.35	Quench harden from 760/780 °C in water. Temper at 150 to 300 °C as appropriate.	—	—	—	—	800	Taps, screwing dies, wood drills, press tools, hand (fitting) tools, files, measuring and marking out in instruments, etc.
	BS 4659: BW1C	1.20	0.35	Quench harden from 760/780 °C in water or oil. Temper at 150 to 300 °C as appropriate.	—	—	—	—	820	Fine edge tools, knives, files, surgical instruments.

Problems

Section A

1. State briefly what is meant by the terms: (i) ferrite; (ii) cementite; (iii) pearlite; (iv) austenite.
2. Since iron is allotropic, state the types of crystal it will form: (i) above 1400 °C; (ii) from 910 °C to 1400 °C; (iii) below 910 °C.
3. State briefly what is meant by the following change points on the iron–carbon thermal equilibrium diagram: (i) the A_1 point; (ii) the A_3 point; (iii) the A_{cm} point.
4. Show by means of a diagram the effect of varying the carbon content of a plain carbon steel on its: (i) hardness; (ii) ductility; (iii) tensile strength.
5. State the essential differences between: (i) wrought iron; (ii) plain carbon steel; (iii) cast iron.

Section B

6. With reference to the iron–carbon equilibrium diagram, explain what is meant by 'recalescence', and describe how this phenomenon may be demonstrated.
7. Sketch the 'steel section' of the iron–carbon equilibrium diagram and describe the cooling from the austenitic condition of: (i) a hyper-eutectoid plain carbon steel; (ii) a hypo-eutectoid plain carbon steel; (iii) a plain carbon steel of eutectic composition.
8. Describe the effects of the following elements on a plain carbon steel: (i) manganese; (ii) silicon; (iii) sulphur; (iv) phosphorous.
9. Describe the approximate composition and properties of the following groups of plain carbon steels. State a typical application in each case: (i) bead mild steel; (ii) mild steel; (iii) medium carbon steel; (iv) high carbon steel.
10. With reference to the steel section of the iron–carbon equilibrium diagram, explain the difference between the A_c points and the A_r points. Explain what is meant by the 'Curie' point.

Chapter 4

The heat treatment of plain carbon steels

4.1 Heat-treatment processes

Plain carbon steels and alloy steels are among the relatively few engineering materials which can be usefully heat-treated in order to vary their mechanical properties. This is because of the structural changes which can take place in the solid iron–carbon alloys. The various heat treatment processes appropriate to plain carbon steels are:

1. annealing;
2. normalising;
3. hardening;
4. tempering;
5. isothermal transformations which take place at a single temperature during a given period of time.

The heat treatment of alloy steels and the use of isothermal transformations will be considered at *level three*. In all the other processes, as applied to plain carbon steels, the steel is heated slowly to the appropriate temperature for its carbon content and then cooled. It is the *rate of cooling* which determines the ultimate structure and properties the steel will have, providing the initial heating has been slow enough for the steel to have reached structural equilibrium at its process temperature.

Before describing the various heat-treatment processes associated with plain carbon steels, a few basic principles need to be revised.

Recrystallisation

During cold-working processes, the grain of the metal becomes distorted and internal stresses are introduced into the metal. If the temperature of the metal is now raised sufficiently, nucleation occurs and

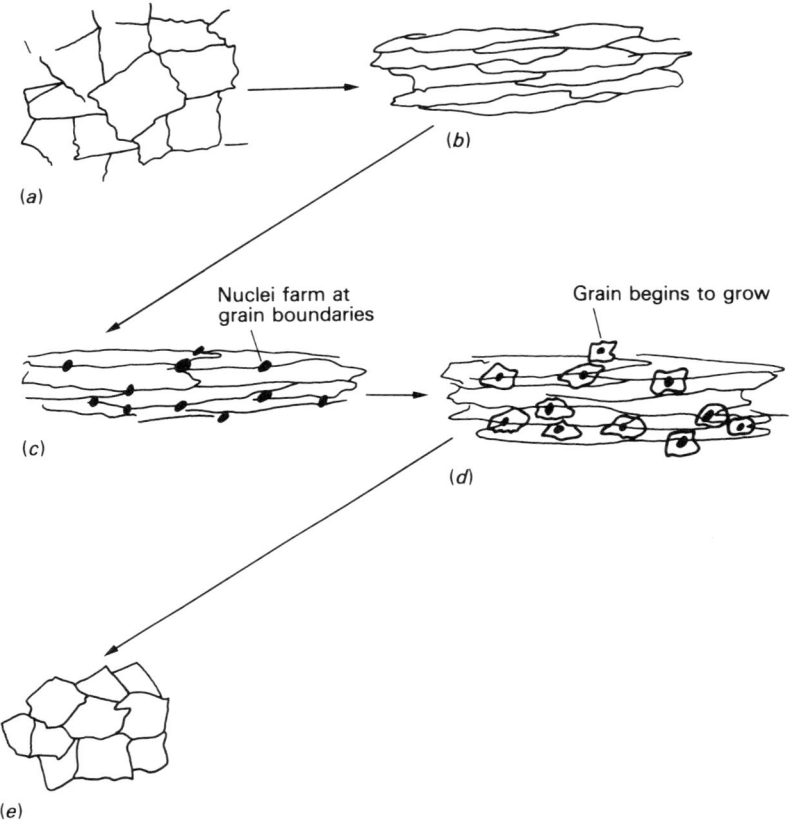

Fig. 4.1 Recrystallisation (*a*) Before working (*b*) After cold-working (*c*) Nucleation commences at recrystallisation temperature (*d*) Grain commences to grow as atoms migrate from the original crystals and attach themselves to the nuclei (*e*) After annealing is complete the grain structure is restored

'seed' crystals form at the grain boundaries at points of maximum internal stress. The more severe the cold-working and the greater the internal stress, the lower will be the temperature at which nucleation occurs for a given metal (see Fig. 4.1). The minimum temperature at which the reformation of the grain occurs is called the *temperature of recrystallisation*. At temperatures above the recrystallisation temperature, the kinetic energy of the atoms on the edges of the distorted grains increases. This allows these edge atoms to move away and attach themselves to the nucleii which then begin to grow into grains. This process continues until all the atoms in the original, distorted crystals have been transferred. Since, after severe cold-working, more nucleii form than the number of original grains, the grain structure after recrystallisation is usually finer than the original grain structure before cold-working.

Cold-working

This occurs when metal is bent, squeezed or stretched to shape *below* the temperature of recrystallisation. Examples of such processes are: pressing out car body panels, drawing rods, wires and tubes, cold-heading rivets, and cold-rolling strip and sheet metal. Cold-working results in severe distortion of the grain of the metal (see Fig. 4.1) and eventually the metal becomes so stiff and brittle that it breaks. (This is what happens in a tensile test). Before it reaches this state the metal needs to be annealed to restore its grain structure, by heating it to just above its recrystallisation temperature, if further cold-work has to be performed upon it.

Hot-working

This occurs when metal is bent, squeezed or stretched to shape *above* the temperature of recrystallisation. Examples of such processes are: forging, hot-rolling, extrusion. Since the process temperature is above the temperature of recrystallisation, the grains reform as fast as they are distorted by the process. If the metal could be retained at this temperature, there would be no limit to the amount of cold-working to which the metal could be subjected. In practice there are strict limitations. The initial temperature has to be limited so that overheating and 'burning' of the metal does not occur or, in some cases, the melting point is not reached. Since the metal cools naturally during processing, the finishing temperature has to be judged so that:

1. It is not too high so that subsequent grain growth occurs.
2. It is not too low so that surface cracking occurs.

Critical temperatures

These are the temperatures at which changes of state (phase changes) occur on thermal equilibrium diagrams. For example, on the iron–carbon diagram the change from austenite to ferrite and pearlite commences, when cooling, at the Ar_3 line and is complete by the time the metal has cooled slowly to the Ar_1 line. Similarly, the change from austenite to pearlite and cementite commences at the Ar_{cm} line and is complete by the time the metal has cooled slowly to the Ar_1 line. The Ar_{cm}, Ar_1, and the Ar_3 lines connect the critical temperatures for the individual alloys of iron plus carbon.

4.2 Annealing processes

All annealing processes are concerned with rendering the steel soft and ductile so that it can be cold-worked or machined. There are three basic annealing processes, and these are:

1. stress-relief annealing at subcritical temperatures (also known as 'process annealing' and 'interstage annealing');
2. spheroidised annealing at sub-critical temperatures;
3. full annealing for forgings and castings.

The process chosen depends upon the carbon content of the steel, its pre-treatment processing, and its subsequent use. Figure 4.1 shows the temperature ranges for the annealing processes superimposed upon the iron–carbon equilibrium diagram. In all annealing processes the cooling rate is as slow as possible and often takes place in the furnace.

4.3   Stress-relief annealing

This process is reserved for steels below 0.4 per cent carbon content. Such steels will not satisfactorily quench-harden (section 4.7) but, since they are relatively ductile, they are frequently cold-worked and become *work-hardened*, that is, the crystal structure is severely deformed from its normal equilibrium condition. Recrystallisation commences at 500 °C but, in practice, annealing is usually carried out at 630 °C to speed up the process and limit grain growth. The principles of hot- and cold-working and recrystallisation were discussed in detail in *Workshop processes and materials: level 1*. To reiterate briefly, hot-working processes are those – such as hot rolling – that are carried out above the temperature of recrystallisation (the temperature at which crystals re-form after being distorted) and crystal distortion is relieved as fast as it occurs. Cold-working processes – such as wire drawing – are carried out below the temperature of recrystallisation, and this results in the structure of the steel becoming severely distorted. Steel in this condition is highly stressed, hard and brittle. The success of stress-relief annealing depends upon the stresses locked up in the crystals triggering off the process of nucleation at sub-critical temperatures. A 'seed' crystal will form at each stress concentration point in the deformed crystal. These seed crystals or 'nuclei' will continue to grow, if the temperature is maintained, until a normal, equilibrium grain structure is restored. Continued heating will result in grain growth and impaired properties.

4.4   Spheroidising annealing

It has already been stated that crystals of pearlite have a laminated structure consisting of alternate layers of cementite and ferrite. When steels containing more than 0.5 per cent carbon are heated to just below the lower critical temperature (650 °C to 700 °C) the cementite in the crystals tends to 'ball up'. This is referred to as the aspheroidisation of pearlitic cementite and the process is shown diagrammatically in Fig. 4.2. Since the temperatures involved are sub-critical, no phase changes take place and the aspheroidisation of the cementite is purely a surface tension effect.

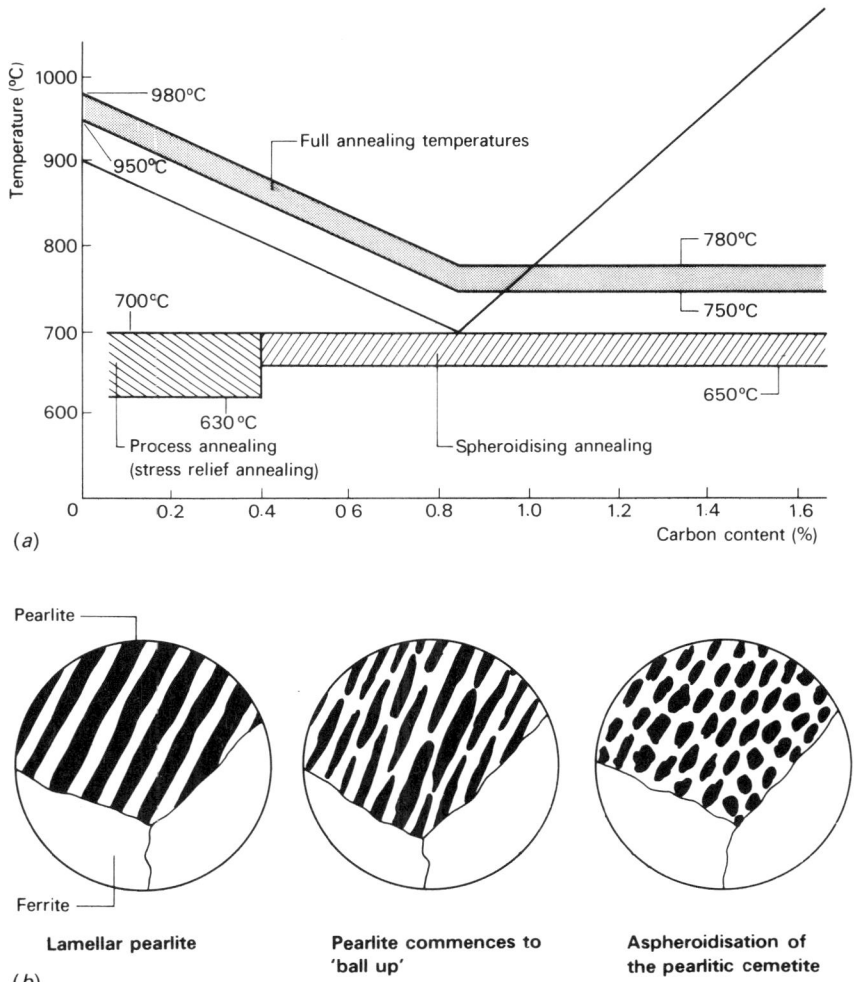

Fig. 4.2 Annealing (*a*) Annealing temperatures (*b*) Spheroidised annealing

If the layers of cementite in the crystal are relatively coarse prior to annealing, they take too long to break down and tend to form coarse globules of cementite. This, in turn, leads to impaired physical properties and poor machined surfaces. Thus, grain refinement by a quench treatment prior to aspheroidisation is recommended to produce fine globules of cementite. The process is most effective when it is used to soften plain carbon tool steels that have been either work- or quench-hardened. After treatment the steel can be drawn and it will also machine freely. Furthermore, steel that has been subjected to spheroidising annealing will harden more uniformly and with less chance of cracking. As with any other annealing process, slow cooling

4.5 Full annealing

Plain carbon steels solidify at temperatures well in excess of the upper critical temperatures with which the heat-treatment processes are concerned, and as a result large castings, well lagged by the sand mould, take a very long time to cool. Similarly, large forgings, although hot-worked below their melting point are, nevertheless, processed at temperatures substantially above their upper critical temperatures for relatively long periods of time. In both cases grain growth is excessive and the physical properties of the metal are impaired. The ferrite settles out along the crystal boundaries of the coarse grains of austenite and also within the grains to provide a mesh-like structure, as shown in Fig. 4.3. This is called a *Widmanstätten structure*.

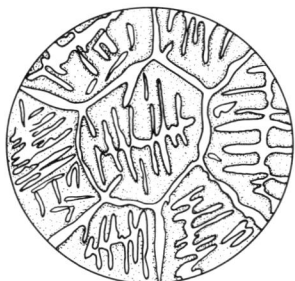

Fig. 4.3 Widmanstätten structure

To render the steel usable it has to be reheated to approximately 50 °C above the upper critical point for hypo-eutectoid steels and to 50 °C above the lower critical point for hyper-eutectoid steels, as shown in Fig. 4.1. This results in the formation of fine grains of austenite, which transform into finer crystals of ferrite and pearlite as the steel is slowly cooled to room temperature, usually in the furance.

4.6 Normalising

The normalising temperatures for plain carbon steels are shown in Fig. 4.4. The process resembles full annealing except that whilst in annealing the cooling rate is deliberately retarded, in normalising the cooling rate is accelerated by taking the work from the furnace and allowing it

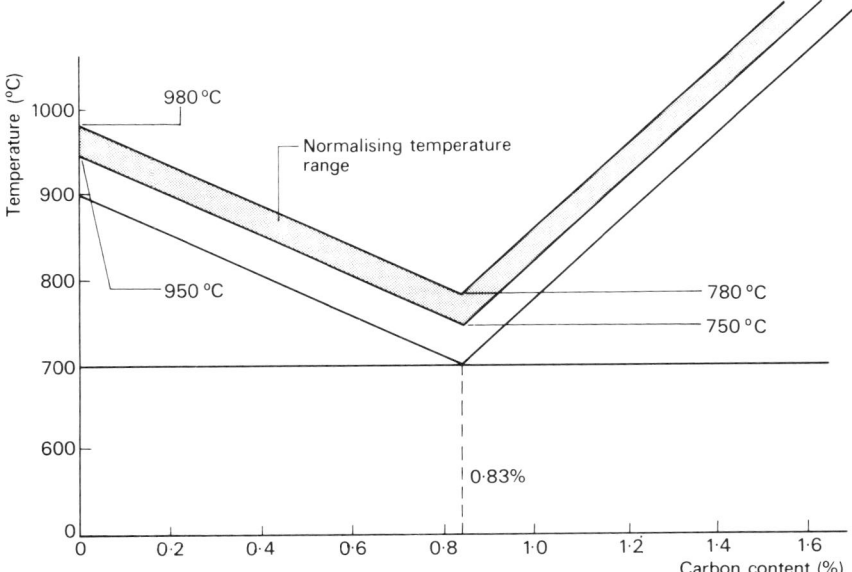

Fig. 4.4 Normalising temperatures

to cool in free air. Provision must be made for the free circulation of cool air, but draughts must be avoided.

In the normalising process, as applied to hypo-eutectoid steels, the metal is heated to not more than 50 °C above the upper critical temperature. Between the lower and the upper critical temperature very fine-grain austenite commences to form. This transformation is completed by 'soaking' the steel at its normalising temperature. By cooling relatively quickly in still air, the fine-grained austenite is converted into fine-grained ferrite and pearlite. The process must be completed as quickly as possible to avoid the grain growth associated with annealing. In the case of hyper-eutectoid steels, fine-grained pearlite and cementite is produced.

The transformation to fine-grained austenite corrects any grain growth that may have occurred previously. The fine-grained structure resulting from normalising improves the strength and toughness of the steel, but reduces its ductility. The reasons for this will be explained in detail at level 3. The increased hardness and reduced ductility allows a better surface finish to be achieved when machining, since excessive ductility leads to local tearing of the machined surface.

However, the ductility is not sufficient for severe cold-working processes. Normalising is frequently used for stress relieving between the rough machining and the finish machining of large components to avoid subsequent 'movement' and loss of accuracy.

4.7 Hardening

If plain carbon steels are quenched (cooled rapidly) from above their upper critical temperatures there is insufficient time for the equilibrium transformations previously described to take place and the steel becomes appreciably harder. The final hardness will depend solely upon the carbon content and the rate of cooling.

The reason for this increase in hardness can be briefly described as follows. When the steel is heated to its hardening temperature it becomes austenitic. However, if it is cooled quickly, the equilibrium transformations into pearlite and ferrite or pearlite and cementite do not have time to take place; crystals of iron carbide just do not have time to form. Therefore as the structure changes from the face-centred crystals (FCC) of austenite to BCC crystals that form below the A_1 temperature, the body-centred crystals (BCC) will be supersaturated with carbon. This distorts the lattice structure to the extent that slip becomes virtually impossible and the metal becomes hard. The solid solution of carbon in BCC crystals that is produced by quenching rapidly from the austenitic condition, produces a structure called *martensite*. This is the hardest structure it is possible to produce in a plain carbon steel and, under the microscope, it appears as acicular (needle-shaped) crystals as shown in Fig. 4.6. Actually these are sections through disc-shaped plates.

It has been stated above that hardness occurs when slip becomes impossible. This is explained in Fig. 4.5. In the annealed condition, metals can be formed by bending, stretching or squeezing them to shape. This is possible because the orderly arrangement of atoms in the crystals allows individual layers of atoms – called *slip planes* – to slide over each other as shown in Fig. 4.5 (*a*). However, if distortion of the lattice occurs (Fig. 4.5 (*b*)) or particles of another material are introduced (Fig. 4.5 (*c*)), then slips cannot occur and the metal becomes hard and brittle. This is explained in greater detail at level 3.

Generally, steels containing less than 0.5 per cent carbon do not harden sufficiently to warrant being considered for cutting tools. Large components do not cool as rapidly as small components and may not achieve the critical cooling rate necessary for maximum hardness. This effect will be considered later in section 4.9.

The critical cooling rate is defined as the slowest cooling rate (quenching rate) from which a steel can be quenched in order to obtain a structure which is wholly martensitic. If this cooling rate is not achieved some fine pearlite will be formed and, although tougher, the steel will be substantially less hard. Exceeding the critical cooling rate will ensure maximum hardness for a given steel. However, there is no virtue in increasing the cooling rate once maximum hardness has been achieved, and more severe quenching will only lead to cracking or distortion of the workpiece.

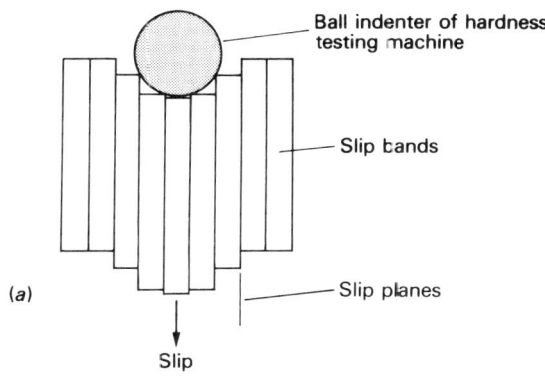

(a)

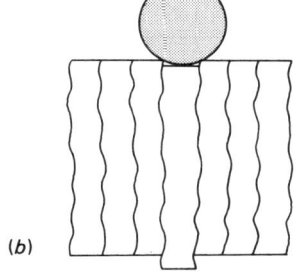

(b)

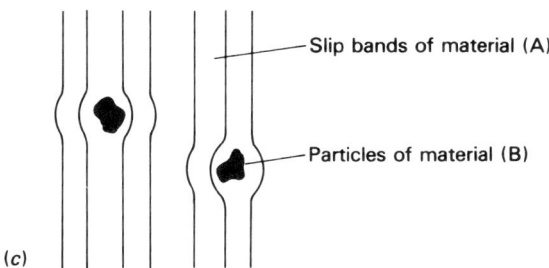

(c)

Fig. 4.5 Slip and hardness (*a*) Indentation is easy in a ductile material as slip occurs. This indicates that the material is soft (*b*) Distortion of the slip planes makes slip extremely difficult. This reduces the amount of indentation indicating that the material is hard (*c*) Particle hardening: the introduction of particles (**B**) distorts the slip planes and makes slip difficult

Under equilibrium conditions the ferrite and cementite form a laminated structure called pearlite. In fact, pearlite can exist in several different forms and the coarsely laminated structure considered so far is called *lamellar pearlite*. Increasing the cooling rate results in the formation of finer and harder pearlite called *sorbitic pearlite*. This is harder and stronger than lamellar pearlite. Increasing the cooling rate further causes the pearlite to 'ball up' into black nodules called *troostitic pearlite* (also called *bainite*). This is harder still. Figure 4.5 shows the appearance of lamellar, sorbitic and troostitic pearlite. A second transformation occurs at 150 °C to 350 °C when the structures shown in Fig. 4.6 (*a*) are changed into *martensite*. Martensite is the hardest structure that can exist in plain carbon steels, and its appearance is shown in Fig. 4.6 (*b*). If the *critical cooling rate* is exceeded the austenite changes directly into martensite without an intermediate transformation into pearlite.

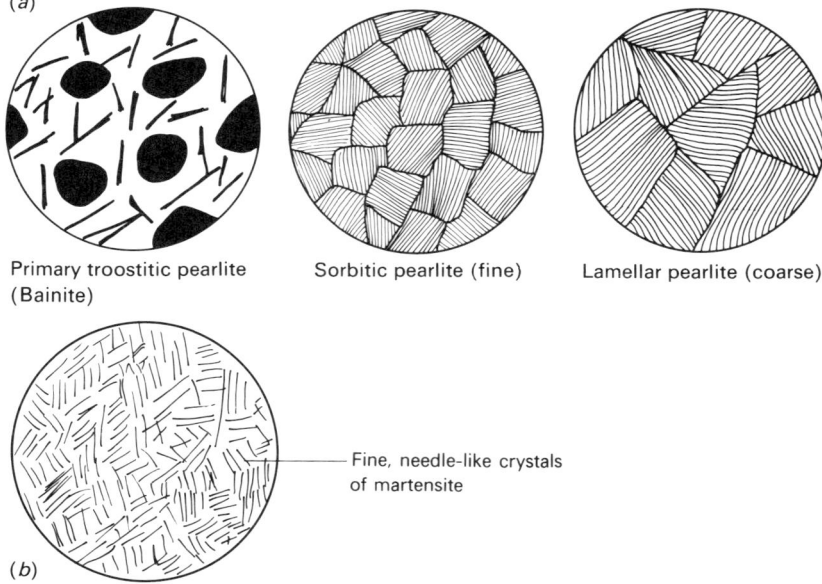

Fig. 4.6 Effects of cooling rate on the structure of plain carbon steels (*a*) Forms of pearlite (*b*) Martensite

There is no particular advantage in heating hyper-eutectoid steels above their upper critical points and, in practice, the hardening temperatures for plain carbon steels are the same as for full annealing. Hardening hyper-eutectoid steels from this lower temperature helps to prevent grain growth, cracking and distortion.

The quenching media is chosen according to the rate at which the steel has to be cooled to give the specified properties. The characteristics of various quenching baths will be considered in detail in Chapter 5, (section 5.17).

4.8 Tempering

A fully-hardened plain carbon steel is brittle and hardening stresses are present. In such condition it is of little practical use and it has to be reheated, or tempered, to relieve the stresses and reduce the brittleness. Tempering causes the transformation of the martensite (see Fig. 4.6) into less brittle structures now to be described. Unfortunately, any increase in toughness is accompanied by some decrease in hardness. Tempering always tends to transform unstable martensite back to the stable pearlite of the equilibrium transformations.

Tempering temperatures below 200 °C only relieve the hardening stresses, but above 220 °C the martensite starts to change into a fine pearlitic structure called troostite. To differentiate the troostite of tempering from the troostite of quench-hardening, the former is called *secondary troostite* (or just 'troostite'), whilst the latter is called *primary troostite* (or 'bainite'). Troostite is much tougher, although somewhat

Table 4.1 Tempering temperatures

Colour*	Equivalent temperature (°C)	Application
Very light straw	220	Scrapers; lathe tools for brass.
Light straw	225	Turning tools; steel-engraving tools.
Pale straw	230	Hammer faces; light lathe tools.
Straw	235	Razors; paper cutters; steel plane blades.
Dark straw	240	Milling cutters; drills; wood-engraving tools.
Dark yellow	245	Boring cutters; reamers; steel-cutting chisels.
Very dark yellow	250	Taps; screw-cutting dies; rock drills.
Yellow-brown	255	Chasers; penknives; hardwood-cutting tools.
Yellowish brown	260	Punches and dies; shear blades; snaps.
Reddish brown	265	Wood-boring tools; stone-cutting tools.
Brown–purple	270	Twist drills.
Light purple	275	Axes; hot setts; surgical instruments.
Full purple	280	Cold chisels and setts.
Dark purple	285	Cold chisels for cast iron.
Very dark purple	290	Cold chisels for iron; needles.
Full blue	295	Circular and band saws for metals; screwdrivers.
Dark blue	300	Spiral springs; wood saws.

* *Appearance of the oxide film that forms on a polished surface of the material as it is heated.*

softer than martensite and is the structure to be found in most carbon–steel cutting tools.

Tempering above 400 °C causes the cementite particles to 'ball up', giving a coarser structure called sorbite. This is tougher and more ductile than troostite and is the structure used for components subjected to shock loads such as crankshafts and connecting-rods in motor-car and motor-cycle engines. It is normal to quench the steel once the tempering temperature has been reached. Table 4.1 gives the tempering temperatures for various applications of hardened plain carbon steels.

4.9 Mass effect

It has already been stated that the hardness of a plain carbon steel depends upon its carbon content and the rate of cooling from the hardening temperature for a given steel. Obviously a thin component will give up its heat and cool more quickly than a thick component if both are quenched in the same quenching bath.

In a thick component the heat will be trapped at the centre so that the core of the component cools more slowly than the outer layers. This leads to a variation in hardness across a section of the component as shown in Fig. 4.7(a). This variation in hardness is referred to as *mass effect*.

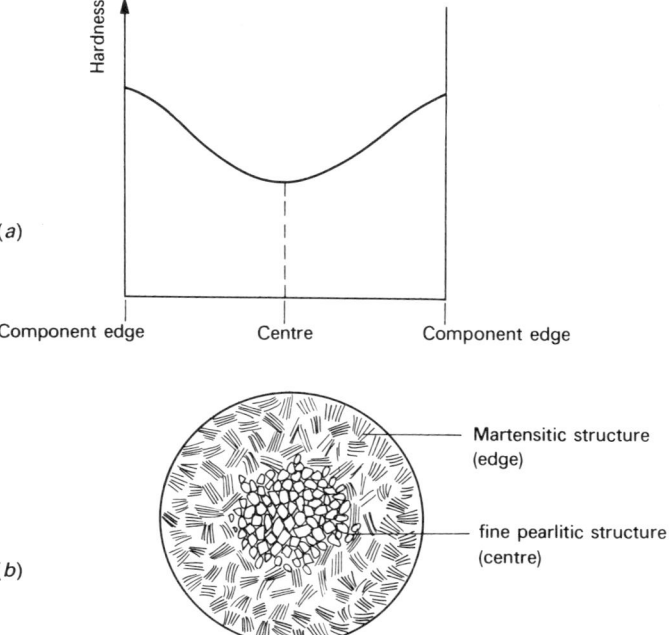

Fig. 4.7 Mass effect (hardenability)

Plain carbon steels have a high critical cooling rate and, therefore, large sections cannot be fully hardened throughout. This is shown in Fig. 4.7 (*b*). Therefore, plain carbon steels are said to have a poor *hardenability*. On the other hand a 3 per cent nickel steel containing only 0.3 per cent carbon will harden uniformly throughout its section because it has a relatively low critical cooling rate. Such a steel is said to have good hardenability. Thus, hardenability can be defined as the ease with which hardness is obtained in a material.

Hardness and hardenability should not be confused. It has already been shown that a 1 per cent carbon steel has poor hardenability compared with a 3 per cent nickel steel containing only 0.3 per cent carbon. However, because of its higher carbon content, the plain carbon steel will exhibit a very much higher surface hardness.

Lack of uniformity of structure and hardness in steels with a poor hardenability can seriously affect their mechanical properties. For this reason it is necessary to specify the maximum diameter or 'ruling section' for which the stated mechanical properties can be achieved under

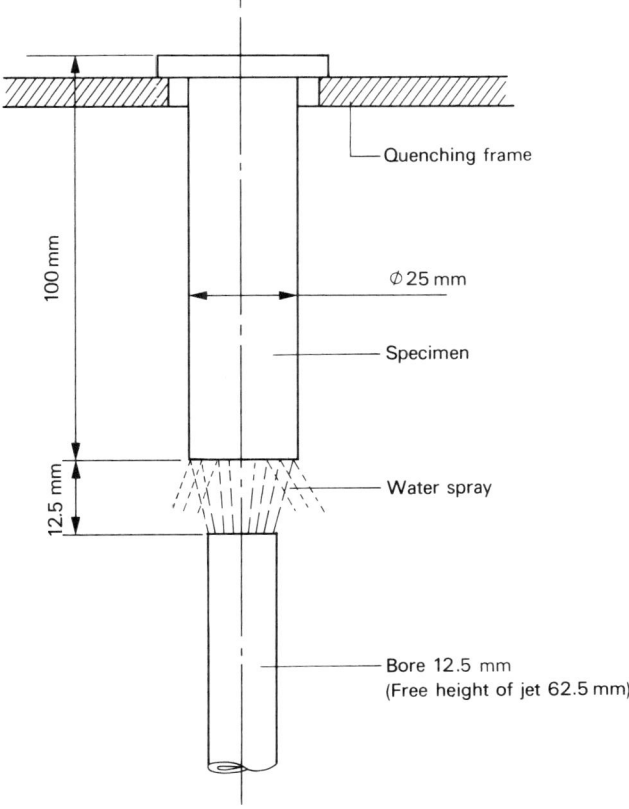

Fig. 4.8 Jominy end-quench test

normal heat-treatment conditions. One of the main reasons for adding alloying elements such as nickel and chromium to steels is to reduce the mass effect and to increase the ruling section for which stated properties can be achieved.

4.10 The Jominy (end-quench) test

This test is used to determine the hardenability of steels. It involves heating a specimen to just above the upper critical temperature so that it is fully austenitic, and then quenching it by spraying a jet of water against its lower end. Details of the test and the specimen are shown in Fig. 4.8.

The specimen cools very rapidly at the quenched end and progressively less rapidly towards the opposite (shouldered) end. A flat is ground along the side of the cold specimen and hardness is tested every 3 mm from the quenched end. The hardness is plotted against distance from the quenched end to give a hardenability curve as shown in Fig. 4.9.

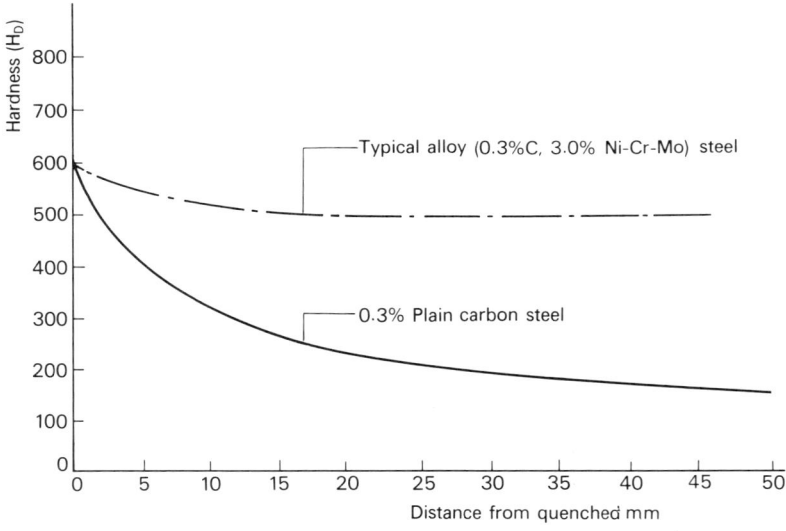

Fig. 4.9 Hardenability curve

4.11 Case hardening

Often components require a hard case to resist wear and a tough core to resist shock loads. These two properties do not exist in one steel. For toughness, the core should not exceed 0.3 per cent carbon content, whilst to give adequate hardness the surface of the component should have a carbon content of approximately 1.0 per cent. The usual solution to this problem is *case hardening*. This is a process by which carbon is added to the surface layers of a low-carbon steel component to a carefully regulated depth, after which the component goes through successive heat-treatment processes to harden the case and refine the core. Thus the process has two distinct steps as shown in Fig. 4.10:

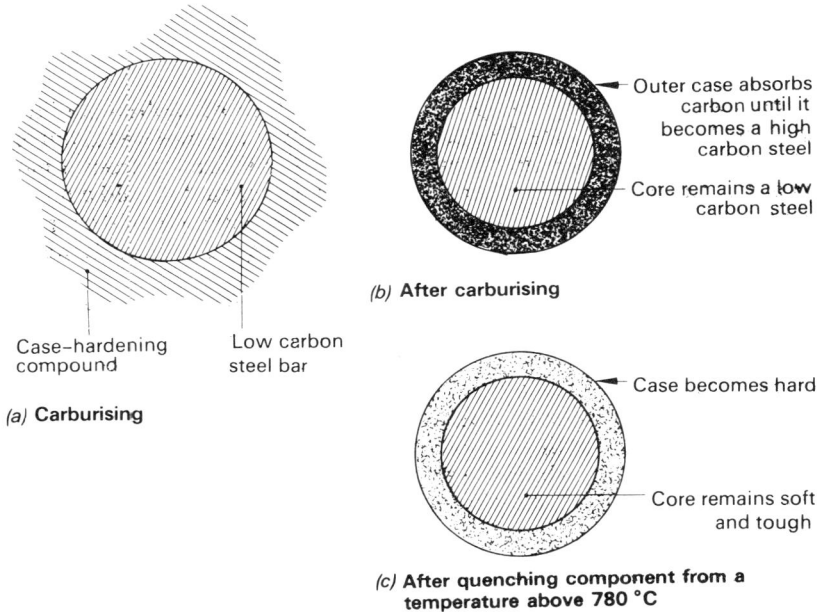

Fig. 4.10 Case hardening

1. *carburising* (the addition of carbon);
2. *heat treatment* (hardening and grain refinement).

Carburising makes use of the fact that low-carbon steels (approximately 0.1 per cent carbon) absorb carbon when heated to the austenitic condition (see Fig. 4.1). Various carbonaceous materials are used in the carburising process.

(a) *Solid media* such as bone charcoal or charred leather, together with an energiser such as sodium and barium carbonates. The energiser makes up to 40 per cent of the total composition.
(b) *Fused salts* such as sodium cyanide, together with sodium carbonate and varying amounts of sodium or barium chloride. Since cyanide is a deadly poison and represents from 20 per cent to 50 per cent of the total furnace content, stringent safety precautions must be taken in its use.

(c) *Gaseous media* are increasingly used now that 'natural' gas (methane) is available. Methane is a hydrocarbon gas containing organic compounds of carbon that are readily absorbed into the steel. Methane is often enriched by the vapours given off by heated oils.

4.12 Pack carburising

This involves packing the components to be carburised into heavy cast-iron or fabricated steel boxes along with the carburising media described in section 4.11. The lid of the box is 'luted' into place with fire-clay to ensure a gas-tight seal. The boxes are then heated to between 900 °C and 950 °C for up to five hours, depending upon the depth of case required. The case depth should not exceed 2 mm, as thick cases tend to flake off due to cracking and brittleness.

When carburising is complete the boxes are allowed to cool down so that they can be opened and unpacked. The components are then cleaned ready for subsequent heat treatment.

4.13 Salt-bath carburising

This involves suspending components in a mixture of molten sodium cyanide and energising salts. The composition of the mixture was given in section 4.11. Large components are suspended individually from a bar lying across the top of the 'pot' or crucible. On no account should copper wire be used as this dissolves in cyanide and the component would drop to the bottom of the pot. Small components are suspended in baskets made from a non-reactive material such as Inconel.

Great care must be taken when using salt-bath furnaces and the safety precautions are dealt with in detail in section 5.18. Cyanide salts have the added hazard of being extremely poisonous. When working with cyanide salts scrupulous personal cleanliness is absolutely essential. Only a few grains of cyanide under the fingernail could prove fatal if transferred to the mouth via food or a cigarette. The process has many advantages over pack carburising for small components and where shallower cases are required, despite its inherent dangers. These are:

1. Loading is quicker and, therefore, cheaper.
2. Heating and carburisation are more uniform with less chance of distortion.
3. The components can be hardened by quenching straight out of the cyanide without the need for further heat treatment.

4.14 Gas carburising

This is carried out in both batch-type and continuous furnaces. The components are heated to 900 to 950 °C in an atmosphere of methane (natural gas) which is a hydrocarbon gas. The gas is often enriched by adding hot mineral oil vapours. Oil is dripped on to a heated platinum electrode. The heat vaporises the oil and the platinum acts as a catalyst, 'cracking' the oil into its constituent elements. The gases are cleaned of moisture and carbon dioxide before being passed into the furnace.

Since such gases are highly flammable, great care must be taken to prevent leakages causing the build-up of explosive mixtures of gas and air.

Gas carburising is used for the mass production of cases up to 1 mm deep.

4.15 Heat treatment after carburising

It is a fallacy to suppose that carburising hardens the steel. It merely adds carbon to the outer layers and leaves the steel in a fully annealed condition with a coarse grain structure. Therefore, additional heat-treatment processes are required to harden the case and refine the grain of both the case and the core in order to give adequate strength and toughness. Reference to Fig. 4.11 will clarify the following descriptions of the hardening and grain-refining processes.

1. *Refining the core.* Since the core has a carbon content of less than 0.3 per cent carbon, the correct annealing temperature is approximately 870°C. After raising the component to this temperature it is water-quenched to ensure a fine grain. The fine grain is required to ensure toughness. Although the temperature of 870 °C is correct for the core (temperature [1] Fig. 4.11) it is excessively high for the case (temperature [2] 4.11).

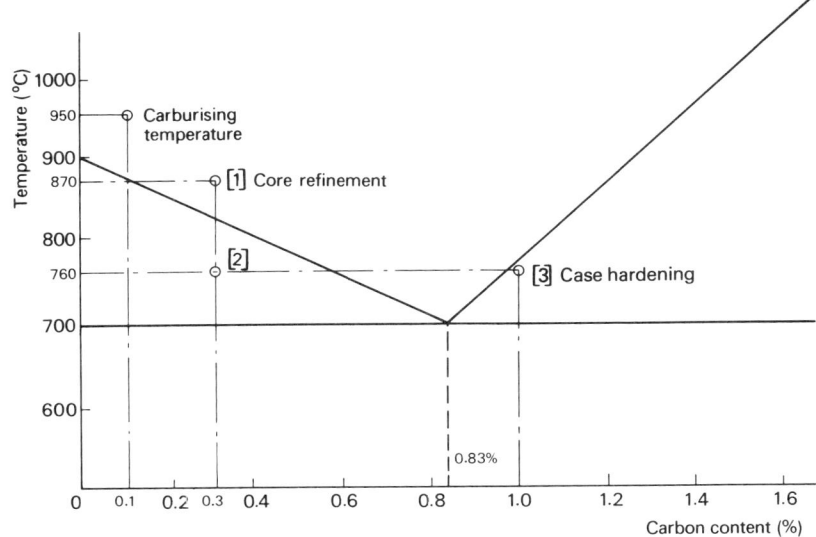

Fig. 4.11 Case-hardening temperatures

2. *Refining and hardening the case.* Since the case has a carbon content of approximately 1.0 per cent carbon its correct hardening temperature is 760 °C. Therefore the component is reheated to this temperature (temperature [3] Fig. 4.11) and again quenched. This hardens the case and ensures a fine grain. The temperature of 760 °C is too low to cause grain growth in the core, providing the component is heated rapidly through the range 650 °C to 750 °C during reheating and quenched without soaking at the hardening temperature.

3. *Tempering.* Tempering at about 200 °C is advisable to relieve any quenching stresses present in the case.

The above procedure is used to give ideal results; however, in the interests of speed and economy the process is often simplified where components are lightly stressed or where alloy steels are used having less critical grain growth and quenching characteristics.

4.16 Localised case hardening

It is often not desirable to harden a component all over. For example it is undesirable to case-harden screw threads. Not only would they be extremely brittle, but any distortion occurring during carburising and hardening could only be corrected by expensive thread-grinding operations.

Various means are available for avoiding the local infusion of carbon during the carburising process. For example:

(*a*) heavily copper plating those areas to be left soft (this cannot be used for salt-bath treatment as copper dissolves in cyanide);
(*b*) encasing the areas to be left soft in fire clay;
(*c*) leaving surplus metal on. This is machined off, together with the infused carbon, between carburising and hardening. Although expensive, this is the surest way of leaving local soft areas (see Fig. 4.12).

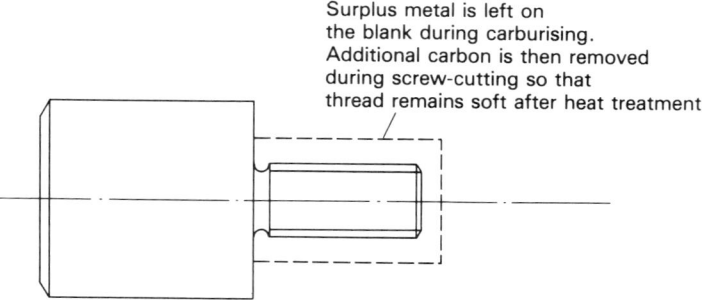

Fig. 4.12 Localised case-hardening

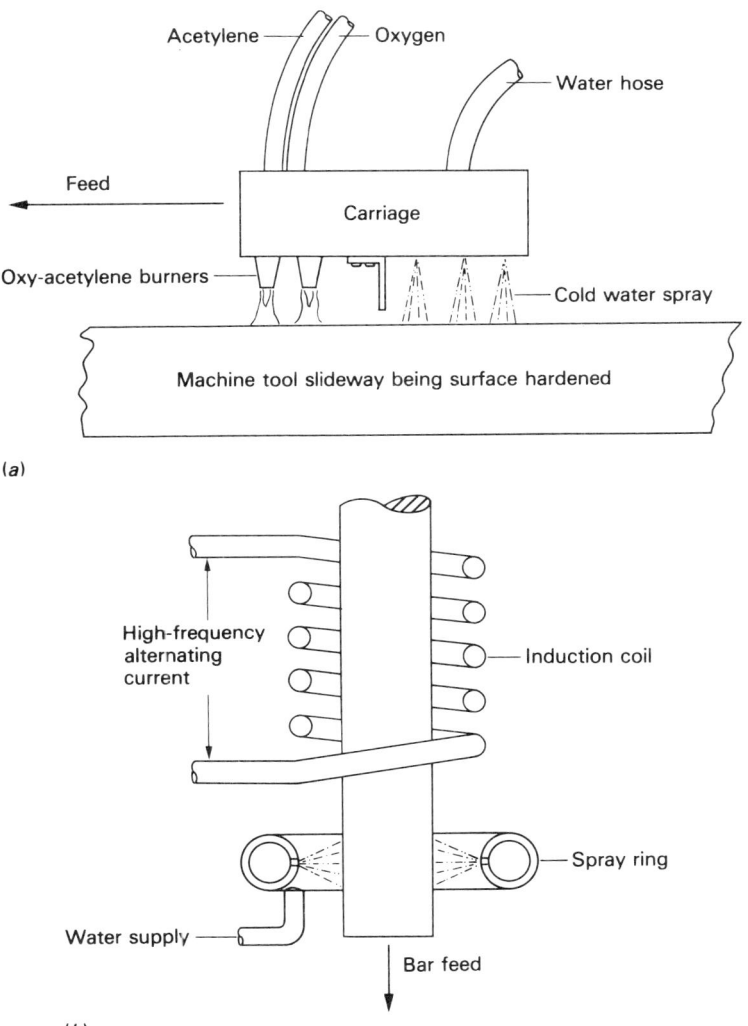

Fig. 4.13 Localised surface-hardening (*a*) Flame hardening (Shorter process) (*b*) Induction hardening

Localised case hardening can also be achieved in medium and high carbon steels by rapid local heating and quenching. Figure 4.13 (*a*) shows the principle of flame hardening. A carriage moves over the component so that the surface is rapidly heated by an oxy-acetylene flame. The same carriage carries the quenching spray. Thus the surface of the component is heated and quenched before its core can rise to the hardening temperature. Figure 4.13 (*b*) shows how the same effect can be produced by high frequency electromagnetic induction. The higher the frequency the shallower the case, since the higher the frequency the nearer to the surface of the component will be the induced eddy currents which cause the heating. The heating coil is often made of tube perforated with fine spray holes so that it can be used for both heating and quenching.

Problems

Section A

1. Explain briefly the meanings of the terms: (i) quench hardening; (ii) tempering; (iii) annealing.
2. Explain briefly how a chisel made from a 0.8 per cent plain carbon steel may be hardened and tempered. How can the heat treatment temperatures be judged visually in this example?
3. Describe the essential difference between full annealing and normalising.
4. Explain briefly why long slender components should be quenched vertically.
5. With reference to the quench hardening of steels, state what is meant by the 'critical cooling rate'.

Section B

6. With the aid of diagrams explain in detail what is meant by the following terms: (i) recrystallisation; (ii) cold-working; (iii) hot-working.
7. Explain in detail the essential differences between: (i) stress-relief (process) annealing; (ii) spheroidising annealing; (iii) full annealing, and give an example where each process would be used.
8. With reference to the iron-carbon equilibrium diagram explain how:
   (a) a hypo-eutectoid plain carbon steel can be hardened and tempered to give maximum toughness. Describe the micro-constituents present after heat treatment.
   (b) a hyper-eutectoid plain carbon steel can be hardened and tempered so that it is suitable for cutting metal. Describe the micro-constituents present after heat treatment.
9. With reference to the iron-carbon equilibrium diagram explain how a mild steel can be case hardened by the pack-carburising process so that it has a hardened and tempered case and a fine-grained, tough core.
10. Explain fully what is meant by 'mass effect' and 'hardenability' as applied to plain carbon steels, and how the hardenability of a steel can be assessed by the Jominy end-quench test.

Chapter 5

Heat-treatment equipment and processes

5.1 Requirements of a heat-treatment furnace

Since the successful heat treatment of a metal depends upon carefully controlled heating and cooling processes, the requirements of a heat-treatment furnace can be as detailed under the headings which follow.

Uniform heating of the charge

The components to be heat-treated may be placed into the furnace singly, in batches, or passed through the furnace on some form of conveyor. Whatever the means of loading the furnace, these components are referred to as the *charge*. Uniform heating of the charge is necessary in order to prevent cracking and distortion of the component due to unequal expansion. It is also necessary to ensure that the physical changes in the metal resulting from the heat-treatment process are uniformly distributed throughout the workpiece.

Accurate temperature control

Since the temperatures involved in heat-treatment processes are critical, not only must heat-treatment furnaces be capable of operating over a wide range of temperatures, but they must be easily and accurately adjustable to the required temperature.

Temperature stability

Not only is it essential that the temperature be accurately adjustable but, once set, the furnace must remain at the required temperature.

This is achieved either by ensuring that the mass of the heated furnace lining (refractory) is very much greater than the mass of the charge, or by providing the furnace with some form of automatic temperature control (metering of the energy input), or by a combination of these expedients.

Atmosphere control

If the furnace charge is heated in the presence of air the surface of the metal becomes heavily scaled. That is, the oxygen in the air reacts with the hot metal of the charge and forms metal oxides on the surface of the components making up the charge. In the case of ferrous metals the carbon content at the surface of the metal may also be oxidised. This is referred to as *decarburisation* of the metal. The combined effects of decarburisation and oxidation are to alter the surface composition of the metal so as to reduce its surface hardness and toughness. Further, these effects can lead to pitting and deterioration of the surface texture of the components so that they may not 'clean up' in any subsequent grinding operation.

To prevent oxidation and decarburisation the air in the furnace is replaced by an inert gas (a gas which does not react with the metal). Alternatively, the charge may be immersed in molten salts heated to the heat-treatment process temperature.

The simplest form of atmosphere control is where the air in the furnace is replaced by the products of combustion of the burnt fuel. However, this is not wholly satisfactory as contamination can occur from the sulphur in the fuel, the nitrogen in the air and any residual oxygen present. Also any hydrogen in the fuel will burn to water vapour which will react with the metal at high temperatures. Any carbon dioxide present will also react with the metal at high temperatures. The formation of metallic sulphides and nitrides causes embrittlement, whilst the presence of water vapour and carbon dioxide causes oxidation and scaling. More sophisticated approaches to atmosphere control are considered in section 5.16.

Economical use of fuel

With world energy costs constantly increasing, the economical use of fuel is essential if heat-treatment costs are to be kept to a minimum. If the furnace can be run continuously on shift work, considerable economies can be made. The fuel required repeatedly to heat the furnace from cold is much greater than that required for continuous running. Furthermore, the continual expansion and contraction of the furnace lining under intermittent usage causes cracking and early failure of the refractory lining. Thus it is more economical for small workshops to contract their heat treatment out to specialist firms who can load their furnaces continuously, than to carry out heat-treatment processes on an occasional, jobbing, basis.

Low maintenance costs

The furnace is lined with a heat-resistant material such as firebrick. Heat-resistant materials are called refractories. Since the furnace must be taken out of commission each time this lining is renewed, it should be designed to last as long as possible. The refractory should also provide the maximum possible heat insulation so as to keep energy losses to a minimum. This is desirable not only for reasons of economy, but also in order to keep the heat-treatment shop as cool and comfortable to work in as possible. The refractory should also have a high specific heat capacity so that it can hold a large reserve of heat energy and help to stabilise the furnace temperature. Unfortunately a refractory lining with a high specific heat capacity also requires a high energy input to bring the furnace up to temperature initially, which is a further reason for running a furnace continuously wherever possible.

In order to keep maintenance costs to a minimum and preserve the refractory lining as long as possible, the furnace should never be run beyond its recommended operating temperature, even for a short length of time.

5.2 Open-hearth furnace

This is the simplest possible form of furnace and its principle of operation is shown in Fig. 5.1. A gas or oil burner plays its flame directly onto the charge, and heat is reflected onto the opposite side of the

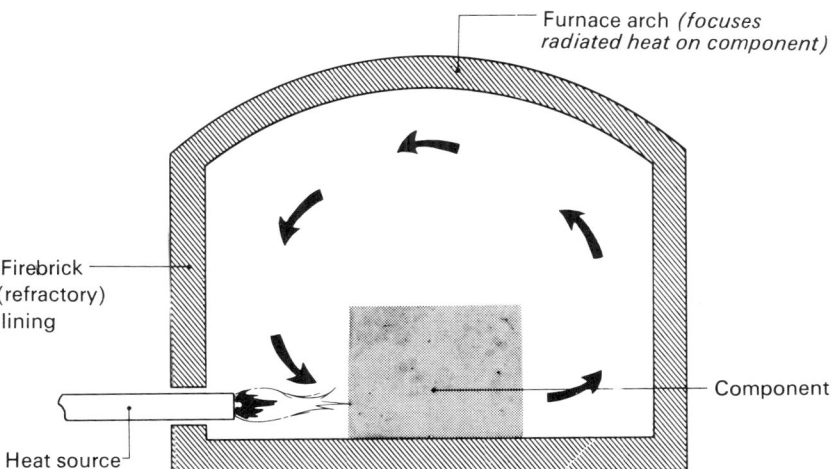

Fig. 5.1 The open-hearth furnace

charge by the furnace lining. The advantages and limitations of this type of furnace are as follows:

Advantages
1. Low initial cost.
2. Simplicity of use and maintenance.
3. Fuel economy.
4. Rapid heating from cold.

Limitations
1. Uneven heating.
2. Poor temperature control.
3. Poor temperature stability.
4. Complete lack of atmosphere control resulting in heavy scaling and contamination of the charge.

5.3 Semi-muffle furnace

The semi-muffle furnace shown in Fig. 5.2 is a considerable improvement upon the open-hearth furnace previously described. The flame from the burner does not play directly onto the charge, but passes under the hearth to provide 'bottom heat'. Bottom heat is provided by

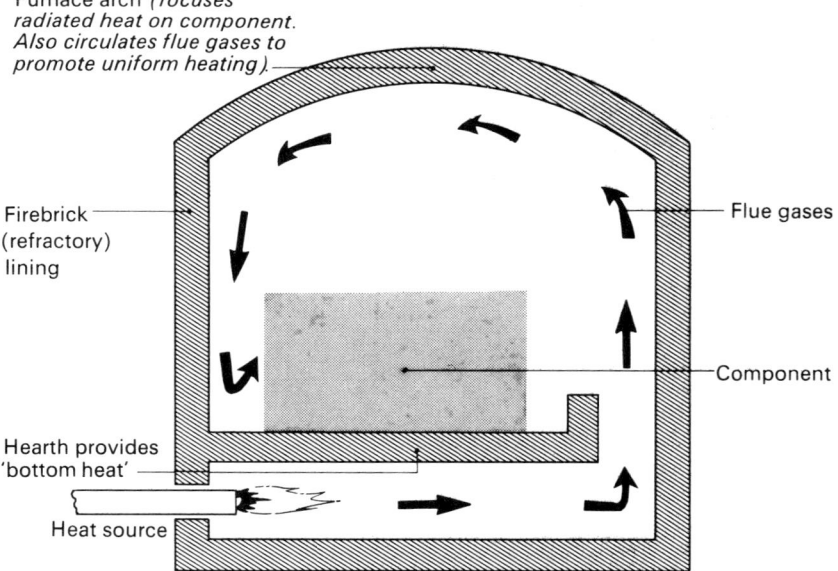

Fig. 5.2 The semi-muffle furnace

conduction and radiation from the hearth, whilst supplementary heating is provided by the circulation of the flue gases and by radiation from the furnace crown. Combustion takes place beneath the hearth and the air supply is carefully controlled so that the flue gases contain a trace of unburnt gas in order to prevent oxidation. The flue gases displace most of the air from the furnace chamber and are drawn off each side of the furnace door so that any air entering at this point is immediately swept up the flue. The advantages and limitations of this type of furnace are as follows:

Advantages
1. Comparatively low initial cost.
2. Simplicity of use and maintenance.
3. Fuel economy.
4. Fairly rapid heating.
5. Heating is more uniform than for the open-hearth type of furnace.
6. Limited atmosphere control by varying the gas/air mixture through a system of dampers.
7. Sufficient temperature control and stability for non-critical applications.

Limitations
1. Heating is still comparatively uneven compared with more sophisticated furnace types.
2. Although oxidation can be reduced by careful control of the gas/air mixture, some scaling will still take place and there will be contamination of the charge by the flue gases.

5.4 Muffle furnace (gas heated)

Figure 5.3 shows a gas-heated muffle furnace. The charge is completely separated from the combustion chamber by an inner chamber or 'muffle'. Thus the conditions for maximum economy of combustion can exist in the combustion chamber, whilst any desired atmosphere can be introduced into the muffle. Obviously, this type of furnace will not heat up as quickly as those previously described and it can only be used efficiently for continuous processing. The advantages and limitations of this type of furnace are as follows:

Advantages
1. Uniform heating.
2. Accurate temperature control.
3. Good temperature stability due to the high mass of refractory material forming the muffle and the furnace lining.
4. Full atmosphere control is possible.

Limitations
1. Higher initial cost.
2. Maintenance more complex and costly.
3. Greater heat loss resulting in lower fuel utilisation efficiency.
4. Not really suitable for intermittent use due to the time taken to heat up the inner chamber.

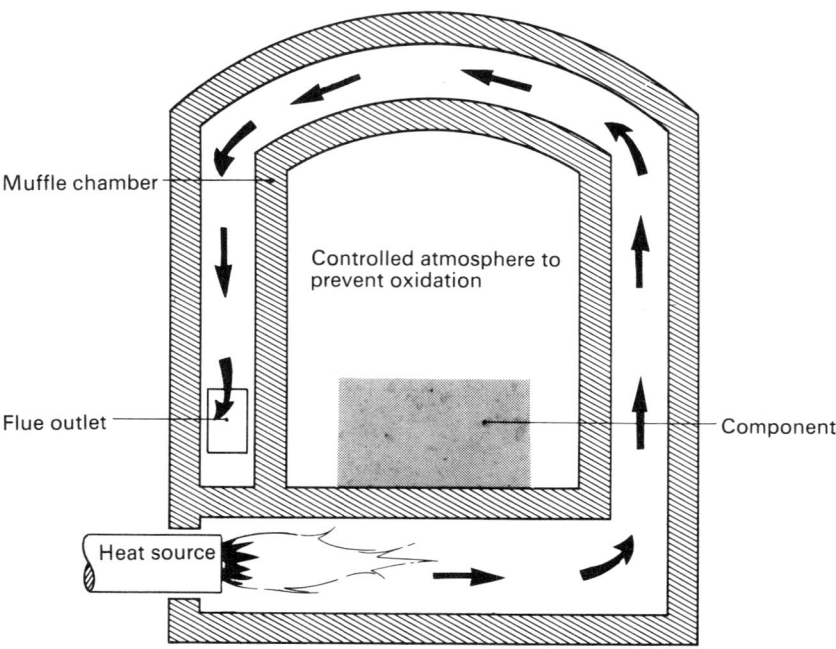

Fig. 5.3 The muffle furnace (gas heated)

5.5 Muffle furnace (electric resistance)

Figure 5.4 shows a typical electric resistance muffle furnace. Since the operation of the electric resistance heating elements is independent of the atmosphere in which they are placed, they may be installed directly into the muffle chamber. Figure 5.5 shows the construction of a typical heating element for a small furnace. Electrical energy is converted to heat energy by a resistance wire element made from an alloy such as 'Nichrome'. For safety, the element is encased in a metal sheath made from a suitable non-corrosive alloy which can withstand high temperatures. The current-carrying resistance wire is insulated from the sheathing by a high-temperature insulating material such as magnesium oxide. Although electricity is a more expensive source of heat energy than gas or oil, the fact that the heating element can be inserted directly into the muffle chamber increases the heating efficiency of the furnace and tends to offset the higher energy cost. Further, electricity is easier to control than gas or oil as a heat source, thus allowing closer temperature control and automatic furnace operation. The advantages and limitations of this type of furnace are as follows:

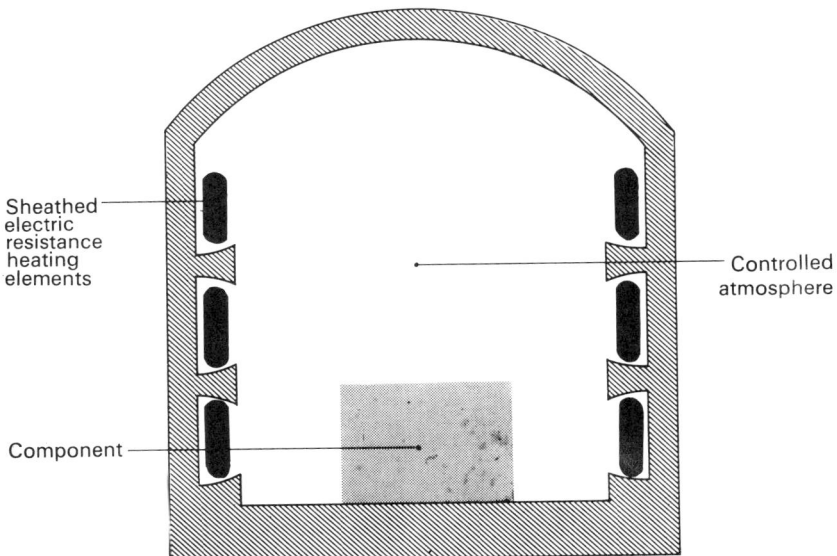

Fig. 5.4 The muffle furnace (electric resistance)

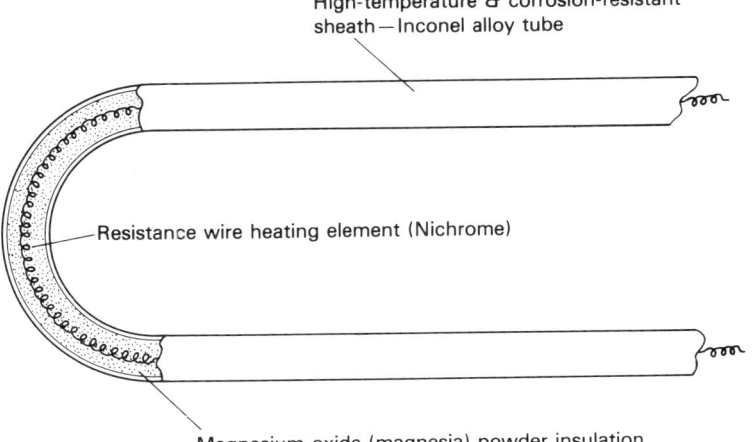

Fig. 5.5 Electric furnace heating element

Advantages
1. Uniform heating.
2. Accurate temperature control.
3. Ease of fitting automatic control systems and instrumentation.
4. High temperature stability.
5. Full atmosphere control.
6. Comparatively easy maintenance.

Limitations
1. Higher energy source costs than for gas and oil.
2. Lower maximum operating temperatures, as above 950 to 1000 °C the life of the resistance elements is substantially reduced.

5.6 Double-chamber furnace

The double-chamber furnace shown in Fig. 5.6 consists of a semi-muffle furnace surmounted by a preheating chamber which is heated by

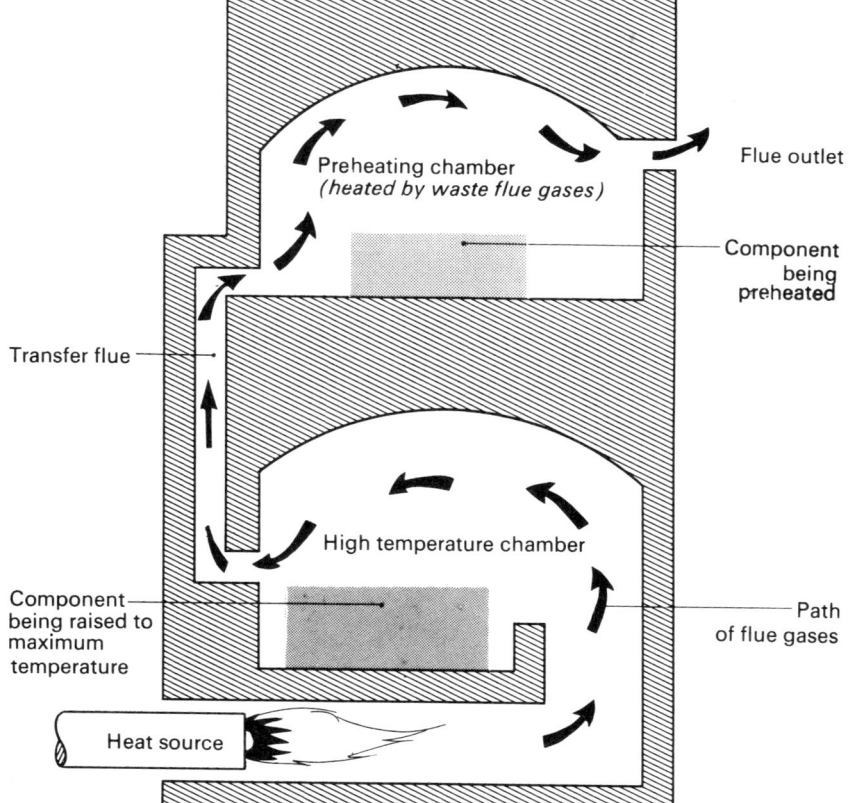

Fig. 5.6 The double-chamber muffle furnace

the waste flue gases from the lower chamber. This type of furnace was once used extensively for the heat treatment of high-speed steel, and it is still used for jobbing work. The high-speed steel components are placed in the preheating chamber and raised in temperature slowly to prevent them cracking. The final heating to the final hardening temperature is then performed rapidly in the lower chamber to prevent grain growth. Once preheated, the components are able to sustain this final rapid heating without cracking.

This type of furnace is now largely superseded by the salt-bath furnace (sections 5.7 to 5.8) for the production hardening of high-speed steels. To obtain the high temperatures associated with the hardening of high-speed steel, forced-draught burners have to be used, the air being supplied to the burners under pressure from a centrifugal blower. The advantages and limitations for the double-chamber furnace are similar to those for the semi-muffle furnace.

5.7 Salt-bath furnace (gas fired)

Figure 5.7 shows a typical gas-fired salt-bath furnace. Points to note are:

1. *Tangential firing* so that the flame does not play directly onto the pot.
2. *Top heat.* In the interests of safety, the salts must be melted from the top downwards. If heated from the bottom, the expanding liquid salts would erupt through the solid crust like a miniature volcano, creating a very dangerous situation, since they would be red-hot.
3. To prevent an explosion throwing the molten salts out of the pot all work and baskets must be dried and preheated.

The salts used are dependent upon the process being carried out, and all reputable suppliers will recommend the most suitable salts to suit a particular situation.

Nitrate-based salts are used for low-temperature applications such as tempering ferrous alloys and the solution treatment of light alloys. If overheated they can cause an explosion and special sections of the Factory Acts cover their use.

Chloride-based salts are used for quench-hardening temperatures of about 800 °C and above.

Cyanide salts are used for case hardening low-carbon steel components. Since these salts are exceptionally and fatally poisonous, extreme care must be exercised in their use and disposal. Special sections of the Factory Acts govern their use.

Note: It is usual to use an 'economiser' in the form of mica flakes floated on the surface of the salts to prevent loss through oxidation and fuming.

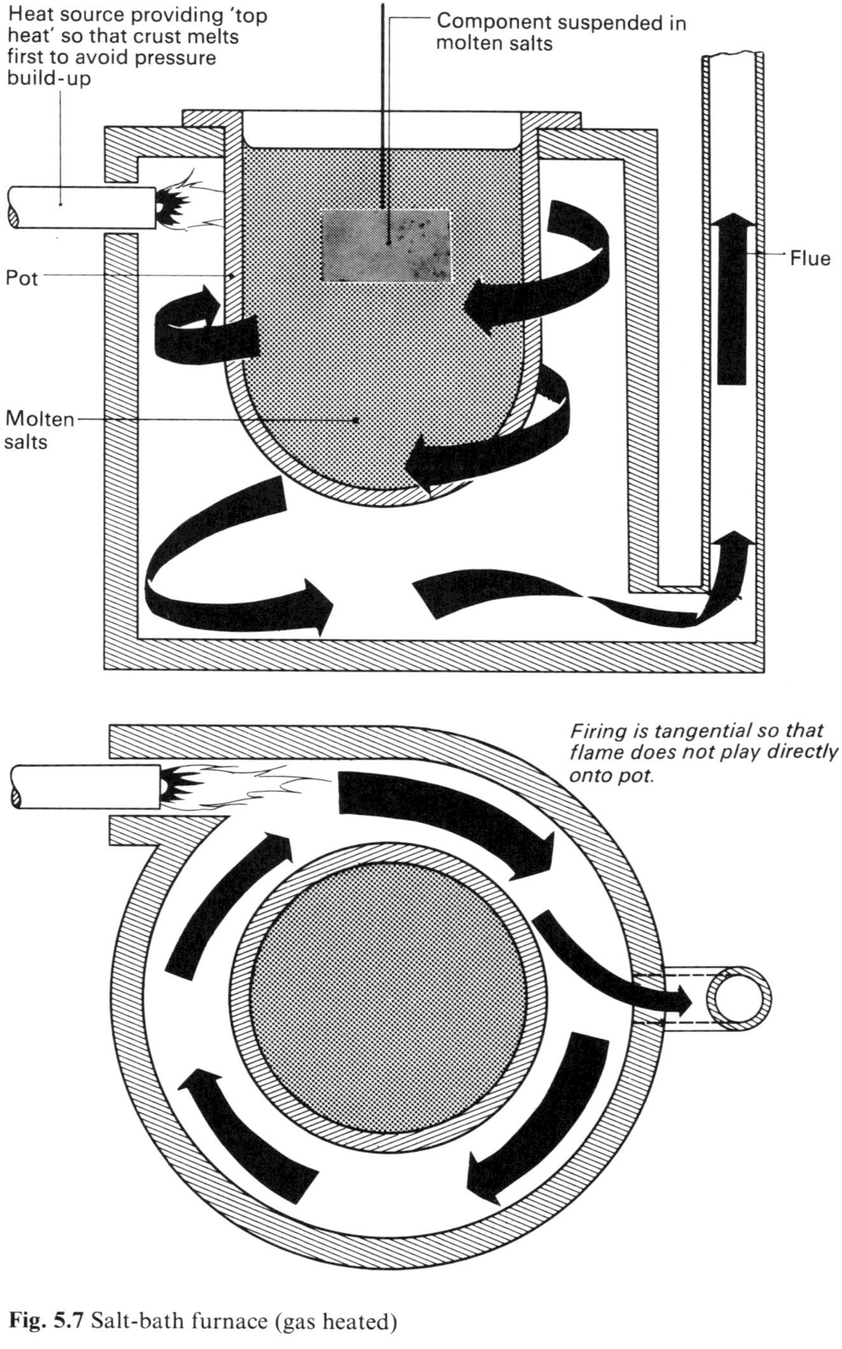

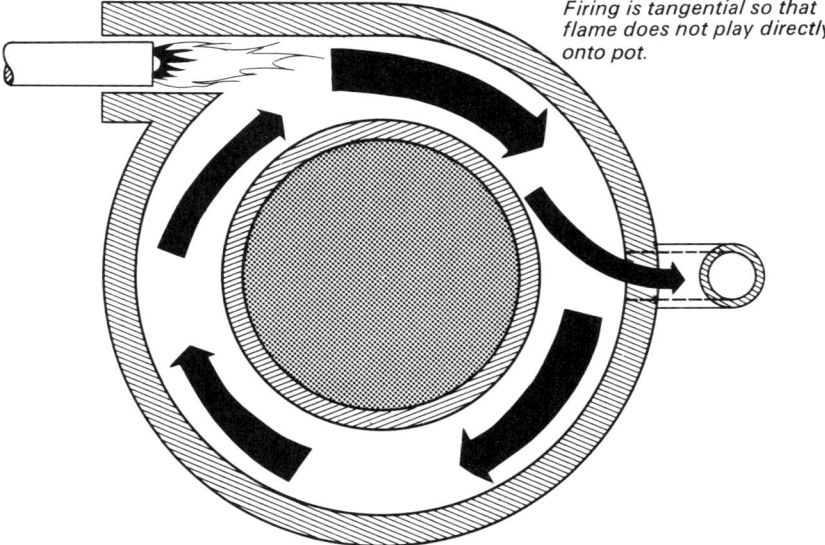

Fig. 5.7 Salt-bath furnace (gas heated)

The advantages and limitations of this type of furnace are as follows:

Advantages
1. Absolute uniformity of heating as the charge is enveloped in molten salt at the treatment temperature.
2. Accurate temperature control.
3. High temperature stability if the mass of molten salts is substantially greater than the mass of the charge.
4. No atmosphere control is required as the charge is enveloped in molten salt.
5. Comparatively simple design and low initial cost.

Limitations
1. Low fuel economy unless run on a continuous-shift basis.
2. Regular maintenance required.
3. Salt baths are potentially dangerous owing to the possibility of eruption of the salts if the work is damp and the risk of an explosion if nitrate salts are overheated. Thus, a relatively highly trained labour force is required to operate salt-bath furnaces.

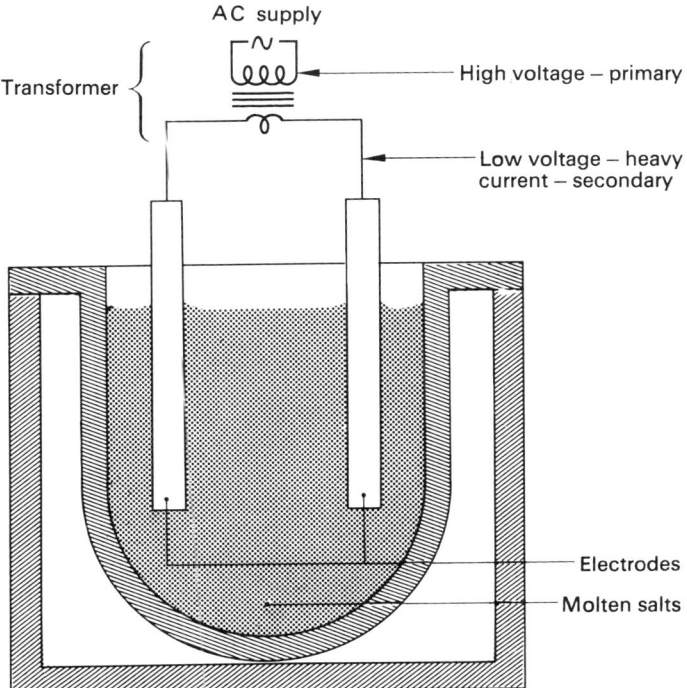

Fig. 5.8 Salt-bath furnace (electrically heated)

5.8 Salt-bath furnaces (electrically heated)

Figure 5.8 shows that in this type of furnace two electrodes are immersed in the salts. These pass a heavy electric current through the salts as a low voltage. The resistance of the salts to the passage of the current causes them to heat up rapidly and become molten. Although electricity is a more expensive energy source than gas or oil, the fact that the heat energy is generated within the salts themselves renders this type of furnace highly efficient and economically comparable with gas- and oil-fired furnaces. Electric heating is readily adaptable to automatic control where high temperature stability is required, as in the heat treatment of alloy tool steels. The general advantages and limitations are the same as for the gas-heated salt-bath furnace.

5.9 Energy sources

Oil, gas and electricity are the three sources of energy used for heating the furnaces described earlier in this chapter.

Oil

This is not widely used except for very large furnaces or where a higher temperature flame than that available from natural gas is required. The advantages and limitations of oil are as follows:

Advantages
1. Cheaper than gas or electricity.
2. Independent of local mains supply.
3. Energy available only limited by the bunkering (storage) capacity of the plant.
4. Higher temperatures are possible than with natural gas.

Limitations
1. Less easily controlled than gas or electricity.
2. Burners are noisy.
3. Burners and pressure feed pumps require regular and skilled maintenance.
4. The products of combustion have an unpleasant smell and tend to contaminate the work.

Gas

This is widely used for heating furnaces of all types and sizes and is available at all industrial centres. The advantages and limitations of gas are as follows:

Advantages
1. Only slightly more expensive than oil.
2. Readily available in all industrial conurbations.

3. Natural gas burns more quietly than oil with a clean flame.
4. Maintenance of the burner equipment is simpler than for oil and is required less frequently.
5. The products of combustion are non-toxic and easily disposable. Their effect on the work is less deleterious than for oil.
6. Gas is readily controllable, both manually and automatically.

Limitations
1. The flame temperature of natural gas is appreciably below that for oil.
2. The energy available depends upon the capacity of local mains.

Electricity

The most expensive form of energy available; it is still widely used because of its controllability and cleanliness. The advantages and limitations of electricity are as follows:

Advantages
1. Temperature control – both manual and automatic – is more precise than is possible with either oil or gas.
2. It lends itself most readily to automatic control.
3. There are no products of combustion to dispose of and no contamination of the work.
4. Maintenance is negligible compared with gas and oil.
5. Although more expensive than oil and gas it can be used more efficiently, especially where atmospheric control is required (see sections 5.5 and 5.8).
6. It lends itself to low-temperature applications such as tempering.

Limitations
1. High cost – although this is to some extent offset by more efficient usage and lower maintenance costs than for gas and oil.
2. Availability limited by the capacity of local mains.
3. Furnace temperature limited to approximately 1000°C if the life of the heating elements is to be reasonable.

5.10 Temperature measurement

The importance of temperature measurement observation and control during heat-treatment processes has already been discussed in Chapter 4. Since errors as little as 20 °C can lead to serious defects in a heat-treated component, accurate temperature-measuring equipment is essential. The familiar mercury-in-glass thermometer is inadequate at the temperatures of most heat-treatment processes. However, mercury-in-steel thermometers and vapour-pressure thermometers can be used for most tempering processes and for the solution treatment of light alloys.

For the higher temperature processes, devices called pyrometers are used to measure the temperature of the furnace and its charge. The more familiar of these are:

(*a*) the thermocouple pyrometer;
(*b*) the radiation pyrometer;
(*c*) the optical pyrometer.

These various temperature-measuring devices will now be considered in detail.

5.11 Mercury-in-steel thermometer

A stainless steel bulb is connected to a pressure gauge by a fine-bore (capillary) tube of steel. The whole system is filled with mercury so that no air or mercury vapour is present. As the temperature of the bulb increases the mercury expands faster than the steel within which it is contained. This increases the pressure of the mercury contained in the system in proportion to the applied temperature. It is this increase in pressure that is registered by the pressure gauge. The pressure gauge is calibrated in degrees of temperature so that direct temperature readings can be made. Providing the connecting tube has a fine bore, and contains a small volume of mercury compared with the bulb and pressure gauge, the connecting tube can be up to 30 m in length. The limiting temperature for this type of thermometer is about 600 °C, at which temperature the mercury commences to vaporise.

5.12 Vapour-pressure thermometers

The basic construction of this type of thermometer is the same as for the mercury-in-steel thermometer described in section 5.11. The difference between the two types of thermometer lies in the fact that in the vapour-pressure thermometer only the bulb is filled with mercury. Any air remaining in the system is evacuated and the capillary tube and pressure gauge is filled with mercury vapour. It is the increase in vapour pressure with temperature that operates the pressure gauge which is calibrated in degrees of temperature. Providing the connecting tube has a fine bore, the bulb and pressure gauge may separate by up to 30 m. The limiting temperature for this type of thermometer is about 800 °C, but it is not so sensitive in the lower ranges as the mercury-in-steel thermometer.

The simplicity and reliability of the mercury-in-steel and the vapour-pressure thermometers make them very attractive for the lower-temperature heat-treatment processes, e.g. tempering.

5.13 The thermocouple pyrometer

This is the most widely used temperature-measuring device for heat-treatment purposes. Figure 5.9 (*a*) shows the principle of the thermocouple pyrometer. If the junction of two dissimilar metal wires – such as iron and copper – forming part of a closed circuit is heated, an electric current will be generated in the circuit. The presence of this

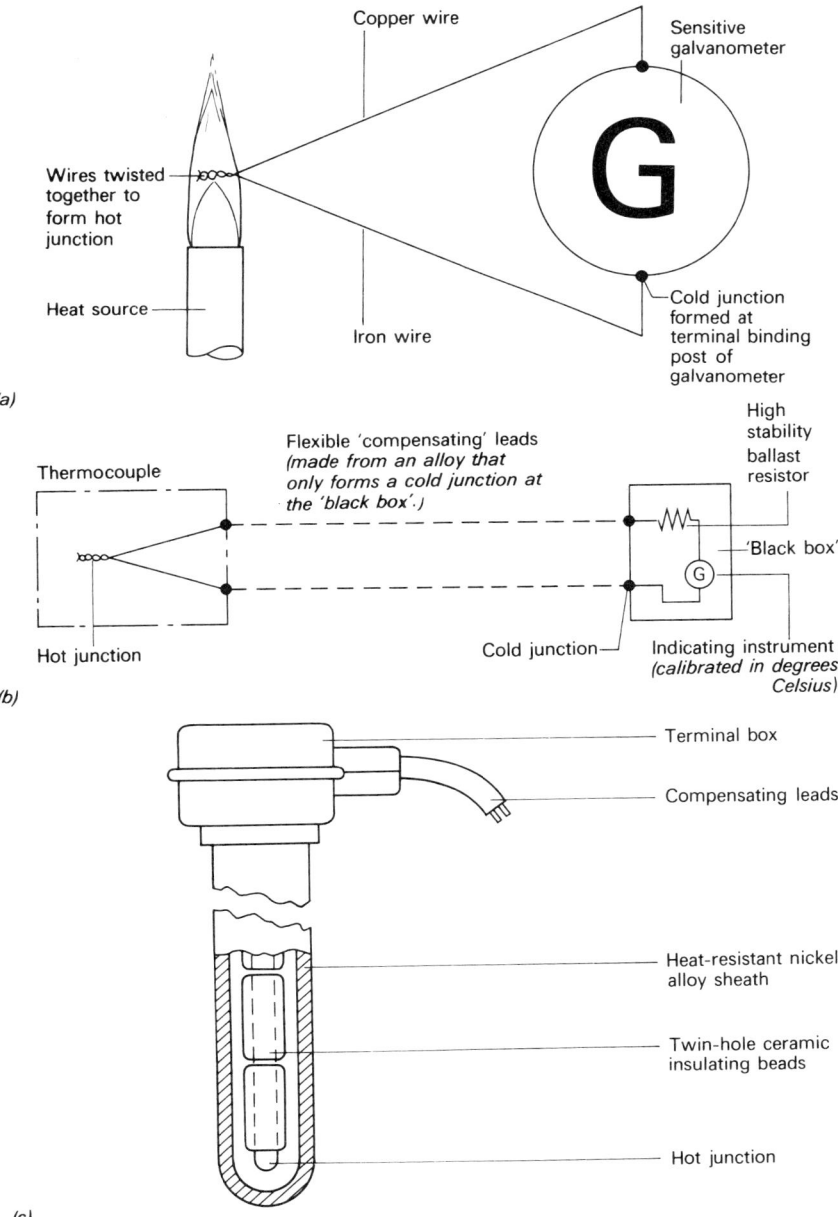

Fig. 5.9 The thermocouple pyrometer (*a*) Principle of operation (*b*) Pyrometer circuit (*c*) Thermocouple probe

current will be indicated by the galvanometer. For a circuit of a given resistance the magnitude of the current flowing will depend upon the potential difference between the hot and cold junctions. The magnitude of this potential difference will, in turn, depend upon:

(a) the metals used to form the hot junction;
(b) the temperature difference between the hot and cold junctions.

Figure 5.9 (b) shows a practical thermocouple pyrometer circuit. The component parts of such a circuit are:

1. the thermocouple probe (hot junction);
2. the indicating instrument (millivoltmeter);
3. the ballast or swamp resistor;
4. the compensating leads.

1. The thermocouple probe. This consists of a junction of two wires of dissimilar metals contained within a tube of refractory metal or of porcelain. Porcelain beads are used to insulate the wires and locate them in the sheath as shown in Fig. 5.9 (c). Table 5.1 lists the more usual hot junction combinations, together with their temperature ranges and sensitivity.

Table 5.1 Thermocouple combinations

Thermocouple	Sensitivity (mV/°C)	Temperature range (°C)
Copper-constantan	0.054	−220 to +300
Iron-constantan	0.054	−220 to +750
Chromel-alumel	0.041	−200 to +1200
Platinum-platinum/rhodium	0.009 5	0 to +1450

Constantan = 60% copper, 40% nickel.
Chromel = 90% nickel, 10% chromium.
Alumel = 95% nickel, 2% aluminium, 3% manganese.
Platinum/rhodium = 90% platinum, 10% rhodium.

2. The indicating instrument. This is a sensitive millivoltmeter calibrated in degrees Celsius so that direct readings can be made. A common error is to set this instrument to zero when the system is installed. In fact, it should be set to the atmospheric temperature at the point of installation. This instrument forms the cold junction and should be placed away from the furnace in as cool a position as possible.

3. The ballast or 'swamp' resistor. This is contained within the case of the indicating instrument, and its purpose is to give stability to the system. The resistance of electrical conductors increases as their temperature increases and the conductors used in a thermocouple pyrometer are no exception. This variation in resistance would seriously affect the accuracy of the calibration of the instrument if the ballast resistor was not present. This resistor is made from manganin wire, and manganin is an alloy whose resistance is unaffected by temperature changes. Making the resistance of the ballast resistor large compared with the resistance of the rest of the circuit swamps the effect of any small changes in resistance that may occur and renders them unimportant.

4. The compensating leads. These are used to connect the thermocouple probe to the indicating instrument. They are made of a special alloy so that they form a cold junction at the indicating instrument, but have no effect when connected to the thermocouple probe terminals. To avoid changes in calibration the compensating leads must not be changed in length nor must alternative conductors be used. The thermocouple, compensating leads and the indicating instrument must always be kept together as a set.

5.14 The radiation pyrometer

The principle of this pyrometer is identical to that described in section 5.13, the sole difference being that in section 5.13 the thermocouple probe was inserted into the furnace atmosphere, whereas in the radiation pyrometer the radiant heat from the furnace or the component is concentrated and reflected onto the thermocouple by a parabolic mirror, as shown in Fig. 5.10.

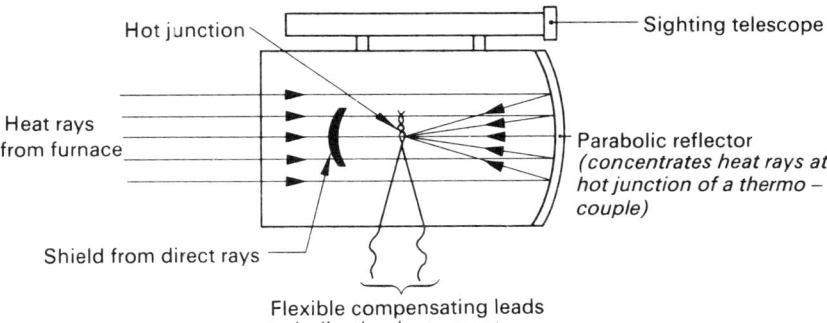

Fig. 5.10 The radiation pyrometer

This type of pyrometer is used in the following circumstances:
1. where the temperature of a large component is to be measured after being removed from the furnace;
2. where the furnace temperature is so high that the thermocouple would be damaged;
3. where the hot component is inaccessible;
4. where the temperature of the component is to be measured in the furnace rather than the temperature of the furnace atmosphere itself.

It must be realised that as the temperature of the component reaches the furnace temperature the rate at which its temperature rises slows down, and it is difficult to assess just when, if ever, the component reaches furnace temperature. Certainly, the soaking time involved would give rise to excessive grain growth. Furnaces are frequently operated above the required temperature, and the component is withdrawn when it has reached the required temperature as measured by the radiation pyrometer.

5.15 The optical pyrometer

The principle of this type of instrument is shown in Fig. 5.11. The brightness of the filament of an electric bulb in the instrument is adjusted until it matches the brightness of the furnace as seen through the eyepiece, at which instance the filament is no longer visible. The higher the temperature of the furnace the brighter it glows.

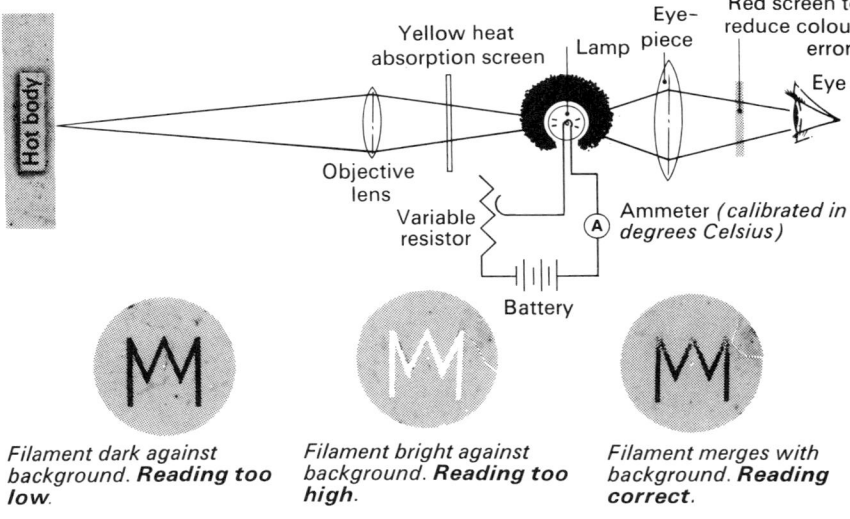

Fig. 5.11 The optical pyrometer

The temperature is read off an ammeter (pre-calibrated in degrees Celsius) measuring the magnitude of the current flowing through the lamp. The main disadvantage of this instrument is that it can only measure temperatures at which a hot body starts to glow – approximately 650 °C in a darkened room. The accuracy of reading is largely dependent upon the operator's skill in judging the precise moment at which the filament merges with its background.

5.16 Atmosphere control

When natural gas is burnt to heat a furnace, excess air is usually present to ensure efficient and complete combustion. The resulting products of combustion (flue gases) will contain oxygen, carbon dioxide, water vapour, sulphur and nitrogen. The oxygen, carbon dioxide, sulphur and water vapour all react with the surface of the metal in the furnace, producing heavy scaling and, in the case of steels, decarburisation of the surface of the component resulting in loss of hardness. The situation is not so serious in muffle furnaces as the fuel is burnt in a separate chamber and cannot come into contact with the work. However, the oxygen and water vapour in the air present in the muffle chamber will still cause some scaling and decarburisation of the work. To eliminate this effect the air in the muffle chamber can be replaced by alternative atmospheres, depending upon the process being performed. This is known as atmosphere control. Atmosphere control is more often associated with production-type heat-treatment furnaces than the simple types of furnace described in this chapter. For general applications exothermic gases (Table 5.2) and endothermic gases (Table 5.3) are used. These atmosphere gases are based upon natural gas (methane) and liquid petroleum gases such as propane and methane. For special applications ammonia and 'cracked' ammonia are used, as listed in Table 5.4. The use of methane for gas carburising atmospheres has already been described in section 4.17. The author is indebted to Wild Barfield Ltd for the information contained in Tables 5.2, 5.3 and 5.4.

5.17 Quenching media

The most commonly used quenching media in order of severity are:

(*a*) compressed air blast (least severe);
(*b*) oil;
(*c*) water;
(*d*) brine (10 per cent solution), (most severe).

The choice of quenching bath depends upon the type of steel being treated and the resultant properties required. Brine (salt and water) is occasionally used to provide very rapid cooling for plain carbon tool steels and case-hardening steels where maximum hardness is required. However, such severe quenching can lead to cracking in all but the simplest components and plain water and quenching oils are most commonly used for both plain carbon and alloy steels.

To avoid cracking and distortion the quenching rate should be no greater than that required to give the required properties in the workpiece. Water provides a quenching rate approximately three times as great as oil and is usually used for plain carbon steels. Oil quenching is usually used with alloy steels as they have a much higher hardenability (section 4.12).

Table 5.2 Exothermic atmosphere gases

Controlled atmosphere	Composition	Source of supply	Heat-treatment processes	Remarks
Exothermic gases ('exogases')	Air/gas ratios: Methane 6/1 to 10/1 Propane 14/1 to 24/1 Butane 20/1 to 32/1 Mixtures of propane and Butane 18/1 to 32/1 Butane 18/1 to 30/1 *Kerosene 500 ft^3/3 pints. Typical chemical composition of exogas obtained from RTG and air.	Methane from supply mains. Propane, butane and mixtures thereof from bottles and tank wagons. Kerosene from liquid supply. *Produces very lean exogas – N_2 84–87% H_2 Nil CO_2 11–13% CO 2–3% CH Nil	Bright annealing of copper and mild and other low carbon steels. Copper brazing of mild steel. Clean hardening of medium carbon steels (using dried gas).	Atmospheres are toxic; combustible to non-combustible, reducing to almost neutral. Exogas can be dried if required. Variations in air/gas ratio permit many variations in the composition of the final exogas. CO_2 and water can be removed from both rich and lean exogas for special applications.

	N_2 (%)	H_2 (%)	CO_2 (%)	CO (%)
Rich exogas	BAL	17	4.0	12
Lean exogas	BAL	2	10.0	2.0

Courtesy of Wild Barfield Furnaces Ltd.

Table 5.3 Endothermic atmosphere gases

Controlled atmosphere	Composition	Source of supply	Heat-treatment processes	Remarks
Endothermic (endogases)	Air/gas ratios: (a) Natural gas 2.2/1 (b) Propane 7.1/1 (c) Butane 9.5/1 (d) Mixtures of propane and butane (Handigas) 8.7/1 Typical analyses: (a) N_2–40%, H_3–40%, CO–20% (b) N_2–45%, H_2–31%, CO–24% (c) N_2–45.5%, H_2–30.5%, CO–24% (d) N_3/45.5%, H_2–30.5%, CO–24%	Methane from mains. Propane, butane and commercial mixtures thereof from bottles and tank wagons. The hydrocarbon gas and air are fed through endo generators.	Clean hardening of carbon steels, alloy carbon steels, etc. Copper brazing of carbon steels, sintering of various metal powders. Carrier gas for gas carburising and carbonitriding.	Toxic, combustible and generally carburising. Gas amenable to automatic dewpoint CO_2 control.

Courtesy of Wild Barfield Furnaces Ltd.

Table 5.4 Ammonia atmosphere gases

Controlled atmosphere	Composition	Source of supply	Heat-treatment processes	Remarks
Anhydrous ammonia	Approximately 100% NH_3	Normally from cylinders containing liquid NH_3 under pressure; net weights 56, 88, 108 and 540 lb. Larger quantity from tank wagons.	Nitriding (a) Normal 'nitralloy' steels – range 490°–530°C. Austenitic steels, range 560°–620° (see also under Carbonitriding). (b) Chrome die steels range 480°–530°. (c) Carbonitriding.	Toxic. All bottles, connections r.h. thread. Batch-type furnaces with retorts; bell-type furnaces.
Cracked ammonia ('cracker gas')	75% H_2 25% N_2 Free ammonia from 0.1 to 0.03% Oxygen Nil H_2O as dewpoint $-40°C$ ($-40°F$)	From ICI Cracker 0–150 ft³/h (type M.I.) 0–800 ft³/h (type H.T.1) fed from cylinders of ammonia.	Bright annealing, etc. of most metals and alloys. Bright tempering and normalising. Reduction of oxides. Sintering.	Cheap supply of pure hydrogen admixed with nitrogen. Protective atmosphere for very high temperature furnaces, molybdenum and tungsten wound.
Burnt ammonia ('ammonia burner gas')	$H_2 = 1$–25% $N_2 = 99$–75% H_2O vapour – either saturation at water-cooler temperature, or, if driers used, saturation at $-40°C$ or lower if special driers employed.	From ammonia burner of suitable size. 15 m³/h 45 m³/h Engelhard Industries	Generally applicable to most bright thermal treatments of most metals and alloys.	Used in all types of furnace.

Courtesy of Wild Barfield Furnaces Ltd.

Air-blast quenching is usually reserved for small sections of high-speed steel where the alloy content is high enough to reduce the transformation rates to a very low level. Sometimes a blast of protective gas (see 'Atmosphere control', section 5.16) is used to give a bright quenched finish to the component.

As soon as the hot workpiece is plunged into the water or oil it becomes surrounded by a blanket of vaporised quenching medium, and cooling can only take place by conduction and radiation through this vapour blanket. Since vapours have low thermal conductivities compared with liquids, the work must be agitated in the quenching bath to disperse the vapour as it forms and keep the work in contact with the liquid. Agitation of the quenching bath also helps to keep its temperature constant and minimises its rise in temperature.

Care must be taken to ensure that distortion is kept to a minimum during quenching, and for this reason it is usual to dip long thin components vertically into the quenching bath. Figure 5.12 shows how cracking and distortion can occur, both by incorrect design and incorrect quenching.

5.18 Safety

There are two main dangers when heat-treating metals:

1. burns that can be both serious and painful;
2. fire.

In properly equipped heat-treatment shops the operators should be provided with adequate protective clothing, including goggles or, better still, a transparent face visor, leather apron and leather gloves.

1. Always wear the protective clothing provided.
2. Assume everything is hot until you have proved it cold.
3. Oil-quenching baths should have an airtight lid. If the oil catches fire, closing the lid stops the air from feeding the flames and puts the fire out. Leave the lid on until the oil cools down.
4. Oil-quenching baths should be provided with a circulating and cooling system to keep the temperature below the flash point of the oil, especially when production quenching large quantities of work. Lubricating oil must *never* be used for quenching, only the special oils developed for quenching are safe and give the desired results.
5. Never light up a furnace until you have been properly instructed and have been given permission.
6. Never use a salt-bath furnace without first drying and preheating everything that is put into the salts.
7. Never tamper with temperature recording and controlling devices.
8. Learn where the fire extinguishers are kept and learn how to use them.
9. Learn what to do if your workmate's clothes catch fire.
10. Learn how to give the alarm if a fire breaks out.

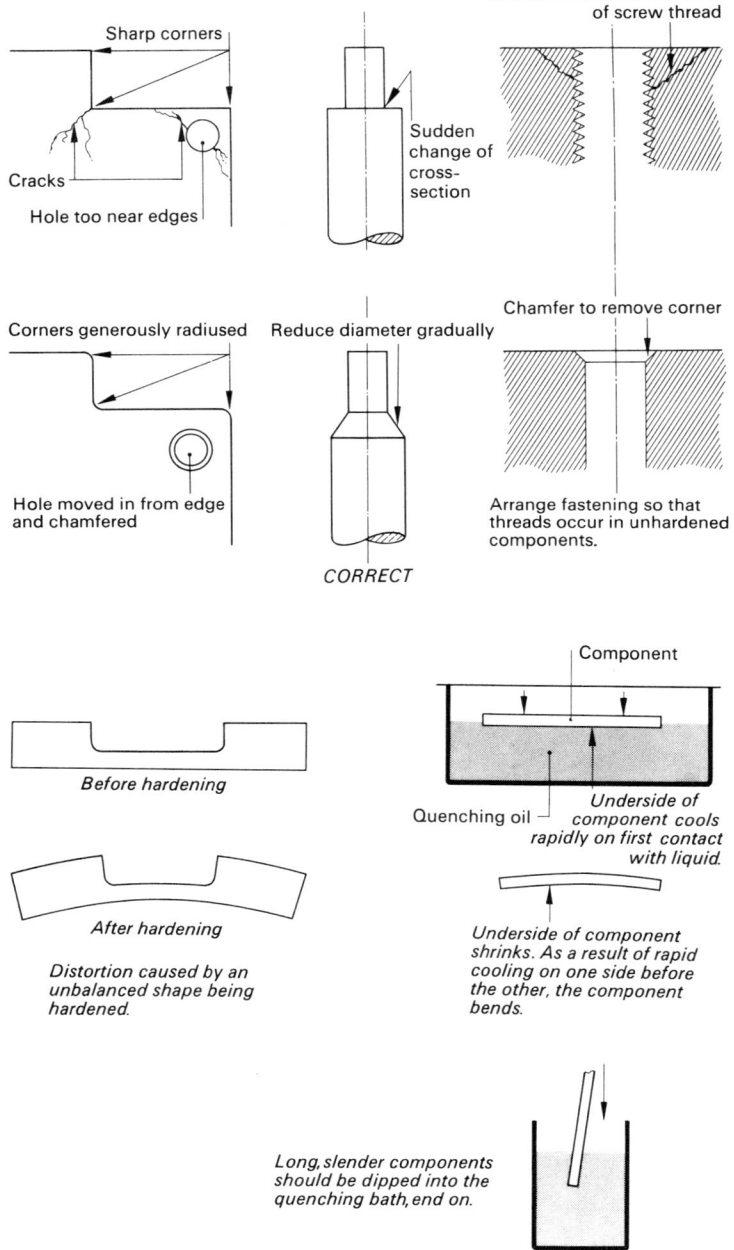

Fig. 5.12 Causes of cracking and distortion

Problems

Section A

1. List *four* essential requirements of a heat-treatment furnace.
2. List the energy sources most widely used for heating heat-treatment furnaces.
3. Describe the safety precautions that should be observed when using: (i) salt-bath furnaces;   (ii) oil-quench baths.
4. State briefly for what purpose the following salt-bath materials would be used: (i) nitrate based salts;   (ii) chloride based salts;   (iii) cyanide based salts.
5. State the normal temperature range for which the following temperature measuring devices would be used: (i) mercury in steel thermometer;   (ii) vapour pressure thermometer;   (iii) iron-constantan thermocouple pyrometer;   (iv) platinum-platinum/rhodium thermocouple pyrometer.

Section B

6. Discuss in detail the requirements of a heat-treatment furnace.
7. Compare and contrast the advantages and limitations of the following heat-treatment furnaces: (i) gas-heated semi-muffle furnace;   (ii) electric-resistance muffle furnace;   (iii) gas heated double-chamber muffle furnace;   (iv) electrically heated salt-bath furnace.
8. (*a*) Explain with the aid of sketches the principle of operation of the thermocouple pyrometer.
   (*b*) Draw a practical thermocouple pyrometer circuit, label the main components and explain their function.
   (*c*) With the aid of a sketch explain how the principle of the thermocouple pyrometer is applied to the radiation pyrometer and explain where such a pyrometer would be used for the heat treatment of metals.
9. (*a*) Compare and contrast the advantages and limitations of the following commonly used quenching media and give practical examples where each would be used: (i) water;   (ii) oil;   (iii) air blast.
   (*b*) Discuss the most common causes of cracking and distortion when hardening steel components and explain how such problems can be minimised.
   (*c*) Discuss the reasons for atmosphere control and the methods by which it can be achieved.
10. Explain with the aid of sketches the principle of operation of the optical pyrometer and discuss its advantages and limitations.

Chapter 6

Common cast irons

6.1 The iron-carbon system for cast irons

Cast iron is the name given to those ferrous metals containing more than 1.7 per cent carbon. It is similar in composition to crude pig-iron as produced by a blast furnace. Unlike steel, it is not subjected to a refinement process. After the pig-iron has been remelted in a cupola furnace ready for casting, selected scrap iron and scrap steel are added to the melt to give the required composition.

Since pig-iron and scrap are cheap, and because there is no refinement process, cast iron is a low-cost material which is useful where a casting of high rigidity, resistance to wear, and high compressive strength is required. Further, cast iron is easy to machine, has a high fluidity which makes it easy to cast into intricate shapes, and has a melting point between 1130 and 1250 °C which is substantially lower than the melting point for mild steel.

Reference back to the iron-carbon thermal equilibrium diagram (Fig. 3.1) shows that the cast irons lie to the right of the 'steel section' of the diagram. Figure 6.1 shows the cast iron section of the equilibrium diagram in greater detail. Since the maximum amount of carbon which can be held in solid solution as austenite is only 1.7 per cent, it is obvious that in all cast irons there will be surplus carbon. This can be taken up by the iron to form cementite, or it can precipitate out as free carbon in the form of *flake graphite*. (Graphite is an allotrope of carbon).

Figure 6.1 shows that there is a eutectic when the carbon present is 4.3 per cent. At this composition the molten metal solidifies at 1147 °C into austenite (γ phase) and cementite. Unless cooling is very rapid,

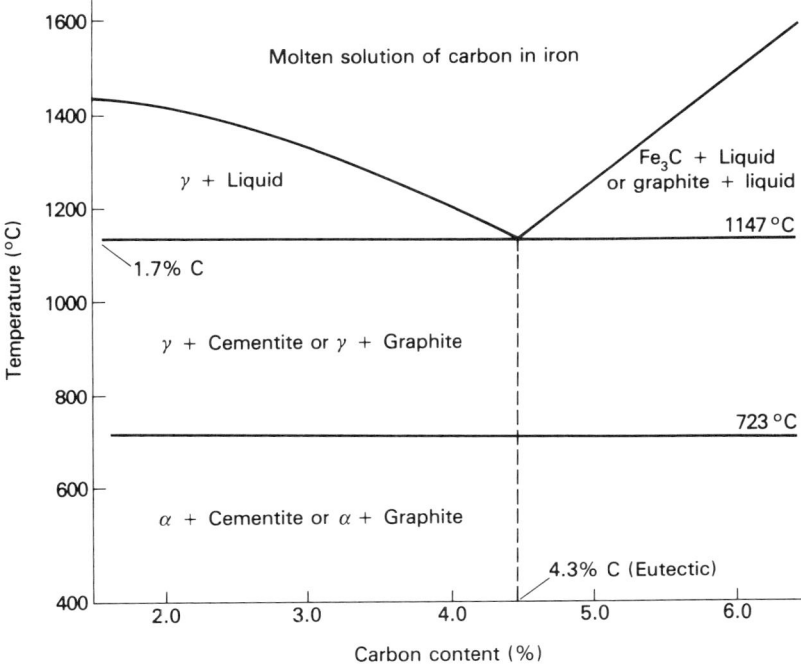

Fig. 6.1 Cast iron section of the iron-carbon thermal equilibrium diagram

graphite may then be precipitated as cooling continues due to the instability of the cementite as a result of some of the impurities present. As cooling proceeds, further graphite is precipitated out from the austenite. At 723 °C the remaining austenite (γ phase) changes into ferrite (α phase). Thus at room temperature the composition will consist of ferrite plus the large flakes of graphite formed on solidification together with fine flake graphite formed by decomposition of the cementite and precipitation from the austenite after solidification.

If cooling is sufficiently fast to prevent thermal equilibrium being achieved, the austenite will change to ferrite and pearlite at the eutectoid temperature of 723 °C. With an even faster rate of cooling the structure will consist of fine flake graphite in a matrix which is entirely pearlite as shown in Fig. 6.2. Ferritic and pearlitic cast irons containing free graphite are called *grey iron* because of the grey appearance of a freshly fractured surface. As in steel, increasing the amount of pearlite present enhances the toughness and hardness of the cast iron.

With even faster cooling a different type of structure is likely to be formed. When the solidus temperature of 1147 °C is reached, the structure will consist entirely of austenite and cementite. The cementite grows due to the precipitation of carbon from the austenite as cooling proceeds, and at 723 °C the remaining austenite changes into pearlite. This type of cast iron is known as *white iron* due to its white appearance when freshly fractured. It derives this appearance from the white crystals of cementite. A microphotograph of a typical white iron structure is shown in Fig. 6.3. The hardness of the cementite in white cast iron makes it difficult to machine and its use is limited mainly to wear-resistant components and as a basic material for conversion to white-heart malleable cast iron (see section 6.4).

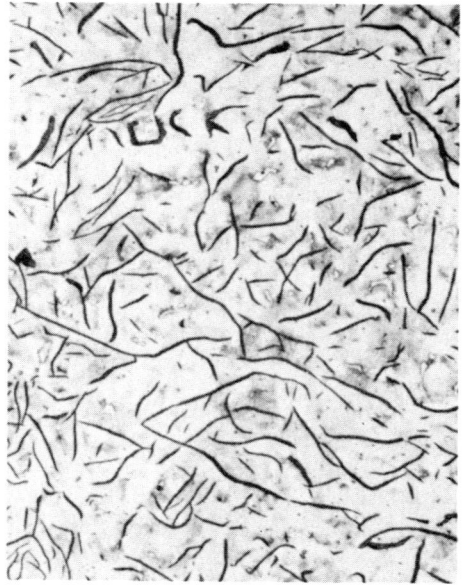

Fig. 6.2 Pearlitic grey cast iron (BCIRA)

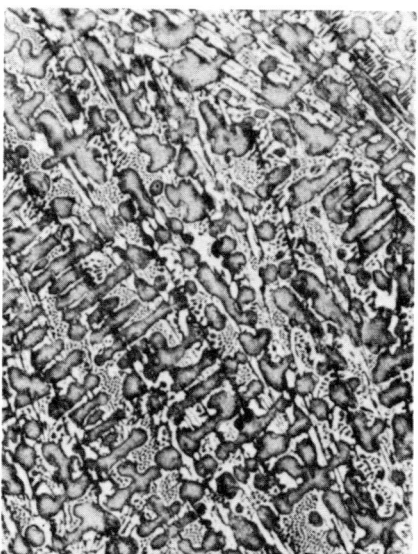

Fig. 6.3 White cast iron (BCIRA)

6.2 Alloying elements and impurities

Cast irons are not just alloys of iron and carbon as the thermal equilibrium diagram would suggest but complex alloys in which impurities such as sulphur and phosphorous and alloying elements such as silicon and manganese have a significant influence on the properties of the cast iron. Although complex alloys, such cast irons are still referred to as common cast irons, the term *alloy cast irons* being reserved for those cast irons containing substantial amounts of such metallic elements as chromium, nickel, etc.

Silicon

This element is used to *soften* cast irons by promoting the formation of flake graphite. The silicon content is increased in irons used for light or thin components that might chill harden by cooling too quickly in the mould and become hard and brittle. The addition of significant amounts of silicon can reduce the eutectic composition down to 3.5 per cent carbon. Thus at a constant rate of cooling, the addition of silicon to a cast iron containing 3 per cent carbon will have the following effects.

(*a*) Ferritic grey cast iron is produced with 3 per cent silicon.
(*b*) Ferritic/pearlitic cast iron is produced with 2 per cent silicon.
(*c*) Pearlitic cast iron is produced with 15 per cent silicon.
(*d*) White cast iron is produced with no silicon.

Providing sufficient silicon has been added to break down the cementite present into flake graphite (3 per cent silicon in this example), there is no benefit to be derived from adding extra silicon. In fact, the presence of excess silicon will lead to increased hardness and brittleness.

Sulphur

The presence of sulphur in cast irons, even in small quantities, has the effect of stabilising the cementite and preventing the formation of flake graphite. Thus sulphur hardens a cast iron. It also causes embrittlement due to the formation of iron sulphide (FeS).

Manganese

The addition of this element in small quantities is essential in all ferrous metals as it combines with any residual sulphur present to form manganese sulphide (MnS). Unlike ferrous sulphide, manganese sulphide is insoluble in molten iron and floats to the top of the melt to join the slag. Thus by removing the sulphur, manganese indirectly softens the cast iron and also removes a source of embrittlement. Excess manganese has the effect of stabilising the cementite and causing hardness in the iron just as sulphur did, but without any embrittlement. Thus it is important to balance the amount of manganese added with great care. Manganese also promotes grain refinement and increases the strength of the cast iron.

Phosphorous

Like sulphur, this is a residual impurity. It is present in cast irons as iron phosphide Fe_3P. This phosphide forms a eutectic with ferrite in grey irons and with ferrite and cementite in white irons. Since these eutectics melt at only 950 °C, high phosphorous irons have great fluidity. Cast irons containing 1 per cent phosphorous are thus very suitable for the production of thin-section castings.

Unfortunately phosphorus, like sulphur, causes embrittlement and hardness in cast irons. Therefore, although a high phosphorus content is desirable in complex, decorative castings, the phosphorus content must be kept low in castings where shock resistance is important.

A typical composition for a grey cast iron could be:

Carbon: 3.3%
Silicon: 1.5%
Manganese: 0.75%
Sulphur: 0.05%
Phosphorus: 0.5%
Iron: remainder

The properties of a typical grey cast iron are summarised in Table 6.1., section 6.7.

6.3 Heat treatment of grey cast iron

It is virtually impossible to arrange for all parts of a casting to cool and solidify at the same rate. It is equally impossible to achieve the optimum cooling rate for a given composition of cast iron in any given casting. Therefore grey iron castings are frequently subjected to heat treatment to relieve stresses, refine the grain structure and improve machineability.

Annealing

Grey irons can be heated to just above the Ac_1 temperature (approximately 760 °C), soaked at that temperature until equilibrium structures have been obtained and then cooled very slowly. The soaking time will depend upon the mass and thickness of the casting. This promotes the precipitation of flake graphite and breaks down any excess cementite which may be forming 'hard-spots'. These hard spots result from over-rapid cooling of thin sections due to the chilling effect of the mould. Annealing also removes internal stresses in the component resulting from uneven cooling in the casting process.

Quench hardening

Ferritic grey iron castings can be heated to just above the Ac_1 temperature and quenched. This results in the formation of pearlite rather than ferrite giving increased toughness and hardness. Great care has to be taken to avoid cracking the castings. For this reason it is not usual to attempt to attain a white iron structure. To relieve the stresses created by quenching it is usual to temper the castings at 450 to 475 °C. Quench hardening grey iron castings is not a common process.

Stress relieving

Most iron castings have internal stresses when they are released from the mould. These stresses not only reduce the strength of the casting, they also cause it to warp and distort during machining. Traditionally, machine beds and frames which require to be dimensionally and geometrically stable were 'weathered'. That is, they were rough machined and then left out of doors for a long period of time varying from months to years. The continual change of temperature causing diffusion of the internal structure, coupled with continual expansion and contraction, reduced the internal stresses and allowed the castings to warp to their final shape. Thus after finish machining no further movement took place.

Unfortunately 'weathering' meant that a lot of money had to be kept tied up in the stock of castings over a long period and this is no longer acceptable. Nowadays castings are stress relieved by soaking them in a furnace at 550 °C (approximately) for a period ranging from several hours to several days depending upon the size of the casting. Slow cooling follows the heating.

6.4 Malleable cast iron

Malleable cast irons are produced from white cast iron by a variety of heat-treatment processes depending upon the final composition and structure required. Malleable cast irons have increased malleability and ductility; increased tensile strength, and increased toughness.

Black-heart process

The castings are heated in airtight boxes out of contact with air at 850 to 950 °C for 50 to 170 hours depending upon the mass and section thickness of the casting. The effect of this prolonged heating is to break down the iron carbide of the white cast iron into small rosettes of graphite. The final structure is of ferrite and fine carbon particles as shown in Fig. 6.4. The name black-heart comes from the darkened appearance of the iron, when fractured, resulting from the formation of free graphite. The relationship of the process to the iron-carbon thermal equilibrium diagram is shown in Fig. 6.5.

White-heart process

The castings are packed in airtight boxes with iron oxide in the form of high grade iron ore. They are then heated to about 1000 °C for between 70 and 100 hours depending upon the mass and section thickness of the

Fig. 6.4 Black-heart malleable cast iron (BCIRA)

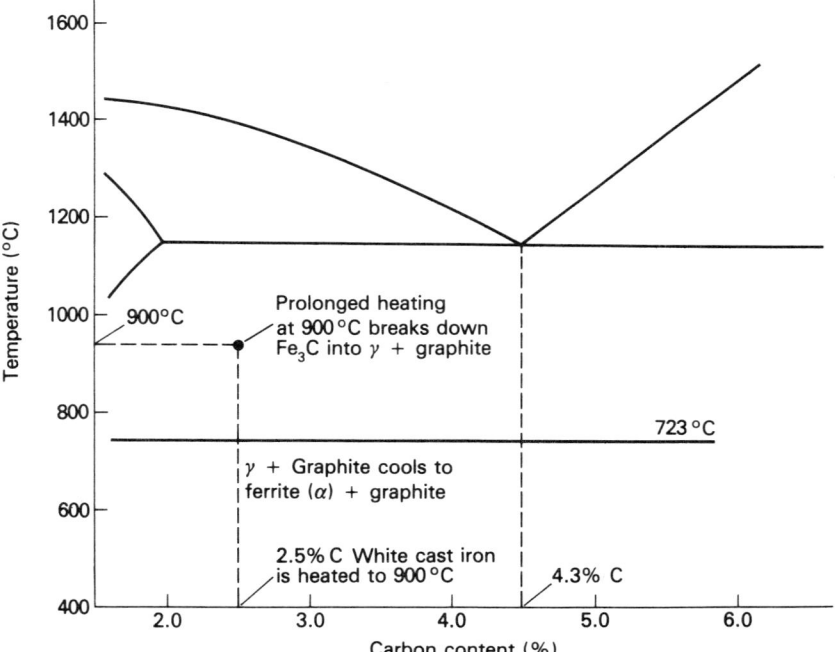

Fig. 6.5 Black-heart transformations

castings. The ore oxidises the carbon in the castings and draws it out, leaving a ferritic structure near the surface and a pearlitic structure near the centre of the casting. White-heart castings behave much as expected of a mild steel casting, but with the advantage of a very much lower melting point and higher fluidity. Figure 6.6 shows a typical white-heart structure.

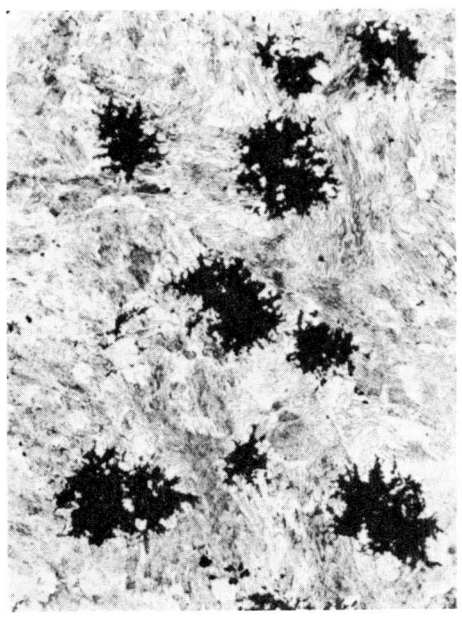

Fig. 6.6 White-heart malleable cast iron (BCIRA)

Pearlitic process

This is similar to the black-heart process inasmuch as the castings are heated in a non-oxidising environment at 850 to 950 °C for 50 to 170 hours. As in the black-heart process the cementite breaks down into austenite and free graphite. However in this process, rapid cooling prevents the austenite changing to ferrite and graphite, and a pearlitic structure is produced instead. This results in pearlitic cast iron being harder, tougher and having a higher tensile strength than is obtained from the black-heart process. However, there is a marked reduction in ductility and malleability.

Pearlitic malleable irons can be produced by increasing the manganese content of the melt to 1.0 to 1.5 per cent. This inhibits the production of free graphite and encourages the formation of cementite and pearlite. Figure 6.7 shows a typical pearlitic cast iron.

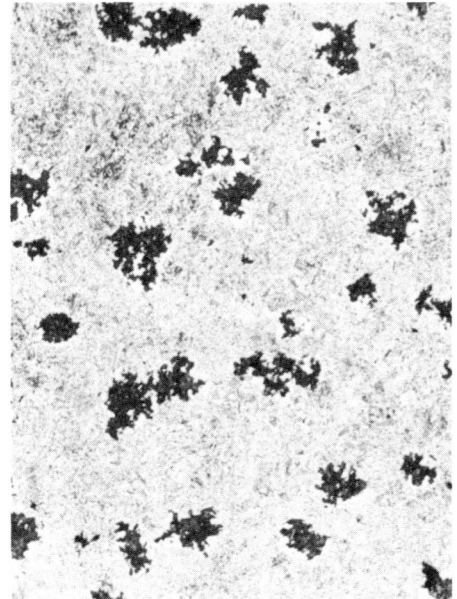

Fig. 6.7 Pearlitic malleable cast iron (BCIRA)

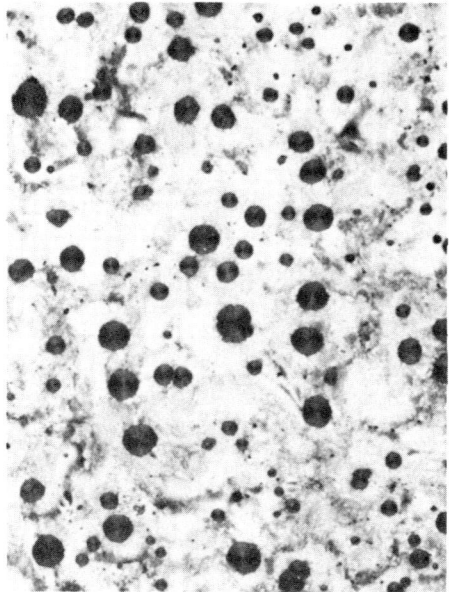

Fig. 6.8 Spheroidal graphite cast iron (BCIRA)

6.5 Spheroidal graphite (S.G.) cast iron

Spheroidal graphite cast iron goes under a variety of names including *nodular iron, ductile iron, high duty iron*, etc. When traces of the metals magnesium or cerium are added to ordinary grey cast iron, the graphite flakes become redistributed throughout the mass of the metal as fine spheroids as shown in Fig. 6.8. It is apparent that the flakes of graphite in a common grey cast iron leave voids in the metal similar to cracks. The rounded nodules of carbon in spheroidal graphite cast iron do not create stress concentrations, and this greatly enhances the strength of the castings. It also reduces the likelihood of fatigue failure.

6.6 Alloy cast irons

The alloying elements in cast irons are similar to those used in alloy steels.

Nickel is used for grain refinement and to promote the formation of free graphite. Thus it toughens the casting.
Chromium stabilises the carbides present and increases the hardness and wear resistance of the casting.
Copper is used very sparingly as it is only slightly soluble in iron. However, it is useful in reducing the effects of atmospheric corrosion.
Vanadium is used in heat-resisting castings as it stabilises the carbides and reduces their tendency to decompose at high temperatures.

6.7 Properties and uses of typical cast irons

White cast iron

This is little used except as a basis for malleable cast irons (section 6.4). It is wear resistant but lacks ductility and breaks easily. It is virtually unmachineable. The composition and properties of a typical white cast iron may be listed as follows:

Carbon	2.5%	Elongation %: Nil	
Silicon	0.8%	Tensile Strength: $250\text{--}450\,\text{N}\,\text{mm}^{-2}$	
Manganese	0.4%	Hardness: $400\,H_B$	
Sulphur	0.08%		
Phosphorus	0.1%		

Grey cast irons

These are the most widely used cast irons and they vary in composition according to specific applications. Table 6.1 lists some typical examples and gives their composition, and uses. The tensile strength of such cast irons varies between 150 and $350\,\text{MN}\,\text{m}^{-2}$.

Table 6.1 Typical grey cast Irons

Composition (%)					Applications
C	Si	Mn	S	P	
3.30	1.90	0.65	0.08	0.15	Motor vehicle brake drums
3.25	2.25	0.65	0.10	0.15	Motor vehicle cylinder blocks
3.25	1.75	0.50	0.10	0.35	Medium machine castings
3.25	1.25	0.50	0.10	0.35	Heavy machine castings
3.60	1.75	0.50	0.10	0.80	Light and medium spun cast water pipes
3.50	2.75	0.50	0.10	0.9	Ornamental castings requiring maximum fluidity but only low strength

Cast irons requiring high *fluidity* so that they can be used for very intricate mouldings are used for decorative iron work, such as architectural tracery, but lack in mechanical strength. High fluidity is obtained by ensuring that the melt has a high silicon content (2.5 to 3.5 per cent) and a high phosphorus content (approximately 1.5 per cent). It is the high phosphorus content which reduces the strength of the metal.

General purpose *engineering* irons have to have reasonable mechanical strength. Ideally they should have fine graphite flakes in a matrix of pearlite. Such a cast iron combines good mechanical properties with good machineability. The silicon content will be dependent upon the thickness of section to be cast. However, it generally varies between about 2.5 per cent for castings having thin sections and 1.5 per cent for castings having thick sections. The phosphorus content must be kept low where shock resistance is important, although up to 0.8 per cent may be present to improve fluidity. Sulphur should be kept as low as possible – 0.1 per cent to avoid segregation, hard spots and embrittlement.

Local hardening of grey iron castings (e.g. slideways) can be achieved by *chilling*. This is done by introducing 'chills' or metal plates into the mould just behind the surface layer of sand to promote rapid cooling and the formation of cementite. The hardness occurs only at the surface of the casting and the core remains grey and tough.

Large heavy castings do not require a high silicon content as there is little danger of chilling. A typical composition would have 1.2 to 1.5 per cent silicon, 0.5 per cent phosphorus, and 0.1 per cent sulphur. Free carbon in the form of flake graphite will be precipitated naturally as such large castings take a very long time to solidify and cool down.

Special cast irons

Table 6.2 lists the properties and applications of typical malleable and spheroidal graphite cast irons and their uses. Malleable irons are much less brittle than ordinary cast irons and are widely used in the auto-

Table 6.2 Special cast irons

Type of cast iron	Condition	Properties			Applications
		U.T.S. (MPa)	Elong^n (%)	Hardness H_B)	
Blackheart malleable	Annealed	290–340	6–12	125–140	Wheel hubs, brake drums, conduit fittings, control levers and pedals.
Whiteheart malleable	Annealed	270–410	3–10	120–180	Wheel hubs, bicycle and motor cycle frame fittings. Gas, water, and steam pipe fittings.
Pearlitic malleable	Normalised	440–570	3–7	140–240	Gears, couplings, camshafts, axle housings, differential housings and components.
Spheroidal graphite (ferritic)	As cast	370–500	7–17	115–215	Water main pipes. Hydraulic cylinders and valve bodies.
Spheroidal graphite (pearlitic)	As cast	600–800	2–3	215–300	Automobile engine, crankshafts and camshafts.

Table 6.3 Typical alloy cast irons

Type of cast iron	Composition (%)							Properties			Applications
	C	Si	Mn	S	P	Ni	Cr	Other Elements	U.T.S. (MPa)	Hardness (H_B)	
Chromidium	3.2	2.1	0.8	0.05	0.17	—	0.32	—	275	230	Cylinder blocks, brake drums and discs, clutch casings, differential carriers, etc.
Wear & shock resistant	2.9	2.1	0.7	0.05	0.10	1.75	0.10	0.8 Mo 0.15 Cu	450	300	Crankshafts for automobile diesel and petrol engines. High strength. Good shock and vibration damping properties.
Ni-hard	2.8	1.3	—	—	—	21.0	2.0	—	—	60	Martensitic iron of great hardness and wear resistance, ore crushing jaws, abrasive material handling components.
Ni-Resist	2.9	2.1	1.0	0.05	0.1	15.0	2.0	6.00 Cu	215	130	A corrosion resistant alloy suitable for valve and pump bodies handling sulphur and chloride solutions.
Silal	2.5	5.0	—	—	—	—	—	—	215	—	High temperature resistance, suitable for exhaust manifolds, furnace components, etc.

mobile industry and agricultural machinery industry for the manufacture of small, stressed components. They are also used in the electrical industry for conduit fittings, switchgear cases and components, etc. Spheroidal graphite cast irons are used for quite highly stressed components in the automobile industry. By designing components with this material in mind, spheroidal graphite castings can, in some instances, replace steel forgings at very much lower cost.

Alloy cast irons
Table 6.3 lists the composition, properties and applications of a selection of typical alloy cast irons.

Problems

Section A
1. State the essential differences between a cast iron and a plain carbon steel.
2. State the essential difference between a grey cast iron and a white cast iron.
3. Name the TWO processes for malleablising cast iron and explain the reason for these names.
4. With the aid of a sketch distinguish between grey cast iron and spheroidal graphite cast iron.
5. Briefly explain the effect of the following alloying elements on a cast iron: (i) Chromium;   (ii) Copper;   (iii) Vanadium.

Section B
6. Discuss in detail the effect of the following elements on a cast iron: (i) silicon;   (ii) sulphur;   (iii) manganese;   (iv) phosphorus.
7. Discuss the effect of the following heat treatment processes on a grey cast iron: (i) annealing;   (ii) quench hardening;   (iii) stress relieving.
8. Describe in detail the following processes as applied to cast irons: (i) black-heart process;   (ii) white-heart process.
9. (*a*) Sketch the cast iron section of the iron-carbon thermal equilibium diagram and label the more important compositions and temperatures.
   (*b*) With reference to the iron-carbon thermal equilibrium diagram distinguish between the formation of ferritic grey cast iron and pearlitic grey cast iron.
10. State the composition of a suitable cast iron for each of the following applications giving reasons for your choice.   (i) A lightly stressed ornamental casting;   (ii) An internal combustion engine crankshaft;   (iii) A machine tool bed;   (iv) A manhole cover;   (v) A mains water pipe;   (vi) A conduit box;   (vii) Brake discs;   (viii) A furnace flue damper.

Chapter 7

Non-ferrous metals and alloys

7.1 Non-ferrous metals

The term 'non-ferrous metals' refers to the thirty-eight metals other than iron that are known to man. The non-ferrous metals that are most commonly used by engineers are listed in Table 7.1. As well as being used as alloying elements, nickel and chromium are also electroplated onto a variety of metals both as a decorative finish and as protection against corrosion.

Two of the most important non-ferrous metals are aluminium and copper. They not only form the bases of many alloys, but they are widely used in their own right as pure metals.

A list of non-ferrous metals is not complete without mention of the new metals listed below. Although known for many years, these metals are new to engineering. It is only since the Second World War that it has been possible to produce these metals in bulk, and it is only recently that there has been a commercial demand for these metals. The new metals are:

1. *Niobium, tantalum, zirconium:* Used for atomic reactor components.
2. *Tellurium:* Used instead of lead in free-cutting alloys.
3. *Titanium:* Used in supersonic aircraft and rockets as it has a higher strength/weight ratio than aluminium and retains its strength at high temperatures.
4. *Beryllium:* Used as an alloying element with copper to make instrument springs. Beryllium copper alloys can be hardened to provide 'non-sparking' tools for use in oilfields and on gas rigs.

These new metals are very expensive compared with the more common engineering materials and are only used where their special properties can be fully exploited.

The pure non-ferrous metals are used mainly where their properties of corrosion resistance and electrical and thermal conductivity can be exploited. They are not widely used for mechanical engineering applications as their mechanical strength is too low. Their mechanical properties are greatly improved by alloying them together.

7.2 Aluminium

Pure aluminium is a weak, ductile metal with a low density, (2.3 g mm^{-3} compared with 7.9 g mm^{-3} for iron). Because of its low strength it is of little use as a structural material, and for such purposes aluminium alloys are used.

High purity aluminium

This is used where corrosion resistance or high electrical conductivity are required. The impurities present in the metal are less than 0.5 per cent. Aluminium has a high affinity for atmospheric oxygen and a film of aluminium oxide quickly forms over any freshly cut surface. This film is virtually homogeneous and prevents further corrosion taking place. It is also virtually transparent, so the aluminium retains its surface appearance for a long time. Unfortunately aluminium reacts violently with alkalis to give off hydrogen, so that care has to be taken not to subject it to caustic degreasing compounds. Neither is it suitable for marine environments. It is used for lining vessels used in the food processing industry, and also for architectural embellishment both internally and externally.

Aluminium is also a good conductor of electricity. It is second only to copper, with a conductivity of approximately two-thirds of that metal. However, because of its low density it is a better conductor when compared on a weight-for-weight basis. For this reason it is used for the cables of the overhead grid system, where the aluminium conductors are laid up over a high tensile steel core to form a composite cable. Aluminium also has a high thermal conductivity, but its use in lightweight heat exchangers is limited due to the difficulties encountered in soldering, brazing and welding it on a production basis.

Commercial purity aluminium

Commercially pure aluminium contains from 1.0 to 0.5 per cent impurities and these have the effect of strengthening the metal at the expense of reducing its corrosion resistance and electrical conductivity. Table 7.2 compares the properties of a typical high purity aluminium with those of a typical commercial purity aluminium. It can be seen from the table that the mechanical properties of the examples chosen are heavily influenced by the amount of *cold-working* that the metal has received.

Table 7.1 Common non-ferrous metals

Metal	Density (kg/m^3)	Melting point (°C)	Properties	Typical uses
Aluminium	2700	660	Lightest of the commonly used metals. High electrical and thermal conductivity. Soft, ductile and low tensile strength 93 MN/m^2	The base of many engineering alloys. Lightweight electrical conductors
Copper	8900	1083	Soft, ductile and low tensile strength 232 MN/m^2. Second only to silver in conductivity, it is much easier to joint by soldering and brazing than aluminium. Corrosion resistant	The base of brass and bronze alloys. It is used extensively for electrical conductors and heat exchangers, such as motor car radiators
Lead	11 300	328	Soft, ductile and very low tensile strength. High corrosion resistance	Electric cable sheaths. The base of 'solder' alloys. The grids for 'accumulator' plates. Lining chemical plant. Added to other metals to make them 'free-cutting'
Silver	10 500	960	Soft, ductile and very low tensile strength. Highest conductivity of any metal	Widely used in electrical and electronic engineering for switch and relay contacts
Tin	7300	232	Resists corrosion	Coats sheet mild steel to give 'tin plate'. Used in soft solders. One of the bases of 'white metal' bearings. An alloying element in bronzes
Zinc	7100	420	Soft ductile and low tensile strength. Corrosion resistant	Used extensively to coat sheet steel to give 'galvanized iron'. The base of die-casting alloys. An alloying element in brass

Metal	Density (kg/m³)	Melting point (°C)	Properties	Typical uses
Chromium	7500	1890	Resists corrosion. Raises strength but lowers ductility of steels. Improves heat-treatment properties	Used as an alloying element in high-strength and corrosion-resistant steels. Used for electro-plating
Cobalt	8900	1495	Improves wear and resistance and 'hot hardness' of high-speed steels	Used as an alloying element in 'super' high-speed steels and in permanent magnet alloys
Manganese	7200	1260	High affinity for oxygen and sulphur. Soft and ductile	Used to de-oxidise steels and to offset the ill-effects of the impurity sulphur. Larger amounts improve wear resistance
Molybdenum	9550	2620	A heavy, heat-resistant metal that alloys readily with other metals	Used as an alloying element in high-strength nickel-chrome steels to improve mechanical and heat-treatment properties. It reduces mass effect and temper-brittleness
Nickel	8900	1458	A strong, tough, corrosion-resistant metal widely used as an alloying element	Used as an alloying element to improve the strength and mechanical properties of steel. Tends to unstabilise the carbon during heat-treatment, and chromium has to be added to counteract this effect in medium and high-carbon steels. Used for electro-plating

Table 7.2 Properties of Aluminium

Type	Condition	U.T.S. (MPa)	Elong^n (%)	Hardness (H_B)
High purity (99.99% Al)	Annealed	45	60	15
	Half-hard	82	24	22
	Full-hard	105	12	30
Commercial purity (99.0% Al)	Annealed	87	43	22
	Half-hard	120	12	35
	Full-hard	150	10	42

NOTE: High-purity aluminium has superior electrical conductivity and corrosion resistance properties.

Table 7.3 Aluminium

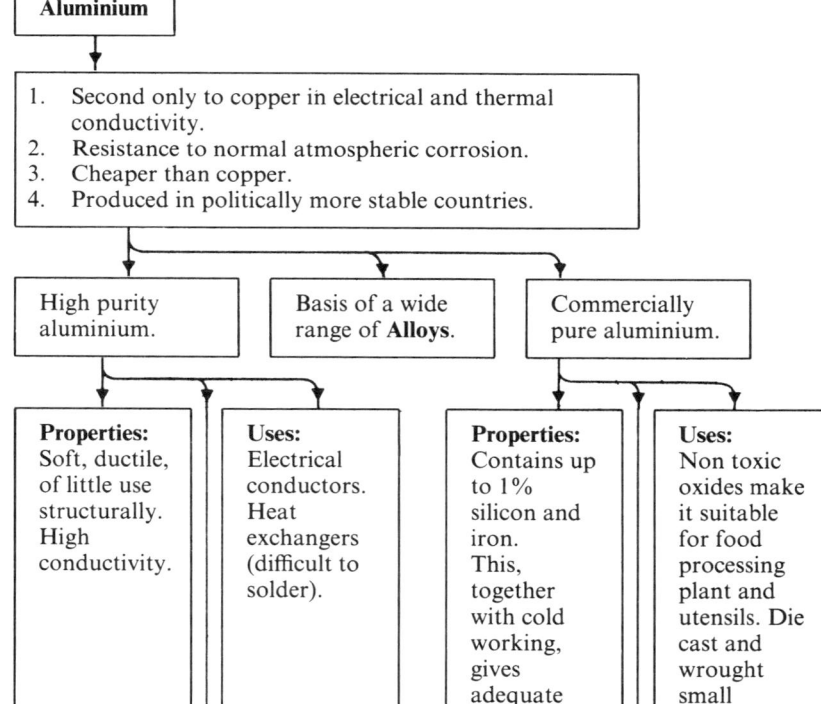

By controlling the amount of cold-working, varying degrees of strength and hardness can be produced. These are said to be the different *tempers* of the metal. Since only a few of the non-ferrous alloys can be hardened by heat-treatment processes, and none of the non-ferrous metals, work-hardening becomes an important consideration.

The properties and uses of aluminium are summarised in Table 7.3.

7.3 Aluminium alloys (non-heat treatable)

Aluminium alloys can be divided into four categories as shown below:

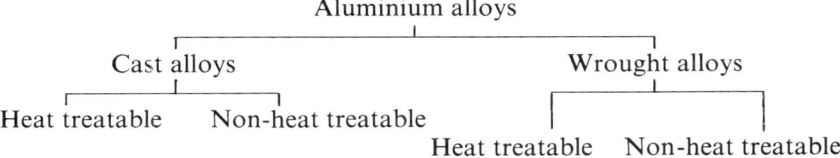

Cast alloys, as the name implies, are those alloys which can be used for producing castings by a variety of processes including sand-casting and die-casting.

Wrought alloys, as the name implies, are those alloys suitable for forming by forging, rolling, extrusion, and drawing.

The non-heat treatable alloys are those which do not respond significantly to heat-treatment processes. As far as the cast alloys are concerned, little can be done other than stress relief and grain refinement treatment. However, the wrought alloys can have their mechanical properties considerably enhanced by a combination of hot- and cold-working.

Casting alloys

These are essentially binary alloys containing aluminium and silicon. It can be seen from the thermal equilibrium diagram shown in Fig. 7.1 (*a*) that silicon is only partially soluble in aluminium below the eutectic composition of 11.5 per cent silicon. Thus the diagram is of the combination type. The addition of silicon to aluminium increases its fluidity and general casting properties.

The microstructure for the eutectic composition shows a coarse eutectic structure of the α phase solid solution plus silicon. In this alloy the α phase is a solid solution of silicon in aluminium. Hyper-eutectic alloys show silicon crystals in a matrix of eutectic structure. This combination of coarse eutectic structure and silicon crystals results in poor mechanical properties and embrittlement, which can be overcome by a process known as *modification*. This consists of adding between 0.005 and 0.15 per cent metallic sodium to the melt immediately before casting. The effect is to delay precipitation of the silicon when the normal eutectic temperature is reached, so that at the commencement of nucleation undercooling results in rapid crystallisation and a fine grain structure with improved mechanical properties. It also raises the eutectic composition to 14 per cent and lowers its temperature to 546 °C, as

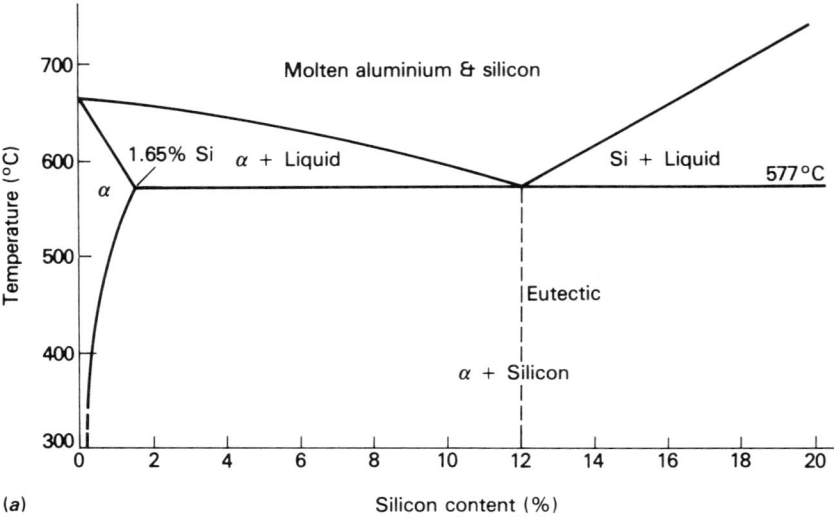

(a)

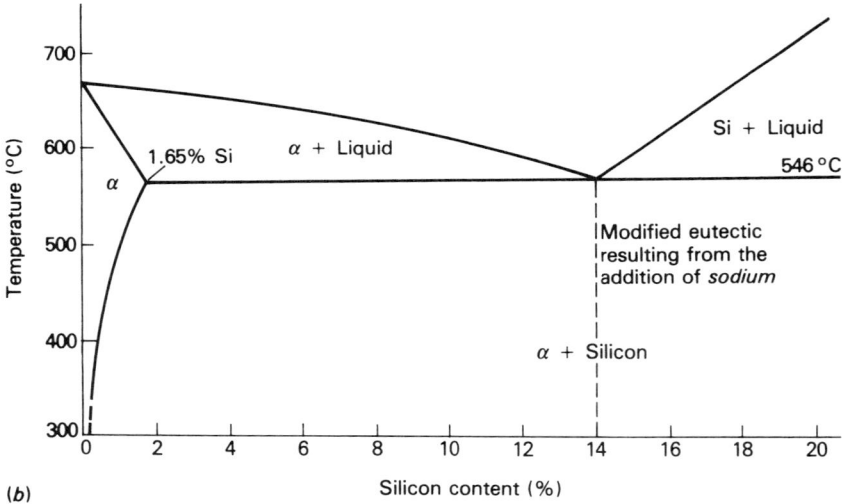

(b)

Fig. 7.1 Aluminium-silicon thermal equilibrium diagrams (a) 'Unmodified' aluminium-silicon thermal equilibrium diagram (b) 'Modified' aluminium-silicon thermal equilibrium diagram

shown in Fig. 7.1 (b). Thus a modified alloy with less than 14 per cent silicon has α phase solid solution crystals in a fine eutectic structure as shown in Fig. 7.2. This results in a casting having increased strength, high ductility and good corrosion resistance. The low melting point and narrow freezing range of alloys only just below the eutectic composition makes the alloy suitable for pressure die-casting.

Fig. 7.2 Aluminium-silicon cast alloy structures (*a*) 13% silicon cast alloy (unmodified) (*b*) 13% silicon cast alloy (modified)

More complex alloys containing aluminium-silicon-copper are used for both sand- and die-casting, whilst aluminium-magnesium-manganese alloys are only suitable for sand-casting. However, these latter alloys are extremely corrosion resistant and are widely used for marine castings. Table 7.4 lists a number of non-heat-treatable aluminium alloys suitable for casting and gives their properties and typical applications.

Wrought alloys

Typical wrought alloys usually contain either traces of manganese or traces of magnesium. A typical non-heat-treatable wrought alloy contains approximately 1 to 1.5 per cent manganese. This increases the tensile strength without significantly affecting the excellent ductility of pure aluminium. This alloy is corrosion-resistant and widely used for kitchen utensils, corrugated sheeting and roof decking panels for the building trade, and drawn aluminium tubing.

Another important non-heat-treatable, wrought aluminium alloy consists of aluminium and magnesium. This alloy is highly corrosion-resistant, particularly to marine environments, and is widely used in shipbuilding. Many ships now use this alloy for their superstructures in order to lower their centre of gravity and make them more stable.

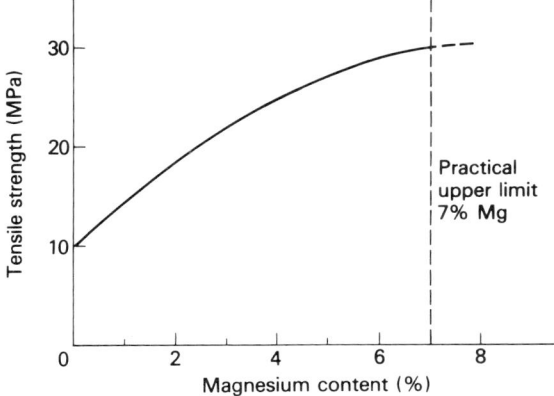

Fig. 7.3 Effect of magnesium content on the tensile strength of annealed aluminium-magnesium alloys

Table 7.4 Non-heat-treatable, cast aluminium alloys

Type	Composition		Condition	Properties			Applications
	Cu	Si		0.1% P.S. (MPa)	U.T.S. (MPa)	Elong^n (%)	
BS 1490/LM2	1.6	10.0	Chill cast	84	224	2.5	Gravity die-casting. General purpose alloy for lightly stressed parts not subjected to mechanical shock.
BS 1490/LM4	3.0	5.0	Sand cast Chill cast	70 80	150 170	2 3	Sand castings; gravity and pressure die castings. General purpose alloy where mechanical properties are of secondary importance.
BS 149/LM6	—	11.5	Sand cast Pressure die-cast	55 85	170 215	7 4	Sand castings; gravity and pressure die castings. Excellent foundry properties. One of the most widely used aluminium alloys when modified. Sumps. Gear boxes, radiators, large castings.
Birmalite	(Mn) (0.35)	10.0	Sand cast Chill cast	98 154	105 168	NIL 0.5	Sand castings and gravity die castings. Maintains its strength. Up to 300 °C, thus used for low-duty pistons and cylinder heads.

Table 7.5 Non-heat-treatable, wrought aluminium alloys

Type	Composition		Condition	Properties			Applications
	Mg	Other Elements		0.1% P.S. (MPa)	U.T.S. (MPa)	Elong^n (%)	
BS 1470/7:N3	(Mn) (1.2)	Cu 0.15 Si 0.6 Fe 0.75	Annealed	45	110	34	Metal boxes, milk bottle caps, food containers, cooking utensils, roofing sheets, panelling of road transport vehicles and railway coaches.
			Hard	170	200	4	
BS 1470/7:N4	2.5	Cu 0.15 Si 0.6 Fe 0.75 Mn 0.5	Annealed	75	185	24	Marine superstructures, lifeboats, panelling subjected to marine environments, chemical plant, panelling for road transport vehicles and railway rolling stock.
			$\frac{3}{4}$ hard	215	265	4	
BS 1470/4:N6	5.0	Cu 0.15 Si 0.6 Fe 0.75 Mn 1.0	Annealed	125	265	18	Ship-building and applications requiring high strength and corrosion resistance.
			$\frac{1}{4}$ hard	215	295	8	
BS 1470/4:N7	7.0	Cu 0.15 Si 0.6 Fe 0.75 Mn 1.0	Extruded	175	308	35	Roofing supports in mines and similar applications requiring high strength and corrosion resistance.

Unfortunately the low melting point of aluminium-magnesium alloys, compared with steel, produces problems where the alloy is used for fire-resistant bulkheads. Figure 7.3 shows the relationship between tensile strength and the magnesium content. It can be seen that the strength increases significantly as the magnesium content is increased up to a practical limit of about 7 per cent. There is no significant reduction in the ductility of this alloy as the strength increases. Table 7.5 lists some examples of non-heat treatable wrought aluminium alloys, together with their composition and typical applications.

7.4 Aluminium alloys (heat treatable)

These are aluminium-copper alloys, together with other elements which respond to heat treatment. The heat treatment of such alloys has already been considered in detail in Chapter 2 when the aluminium-copper thermal equilibrium diagram was discussed. These alloys are annealed (softened) by solution treatment. This consists of quenching the alloy from about 480 °C so as to form a supersaturated solid solution of the α phase. However, this is unstable and the intermetallic compound $CuAl_2$ starts to precipitate out over about four days. This is called artificial ageing, or age hardening. Alternatively the process may be speeded up by heating to about 165 °C for about ten hours. This is called precipitation hardening.

The precipitation process chosen has an appreciable effect on the strength and hardness of the alloy as can be seen from Fig. 7.4. Since intermetallic compounds tend to be hard and brittle, (compared with the softer and more ductile solid solutions), it is the presence of the finely divided particles of $CuAl_2$ throughout the matrix of α phase solid

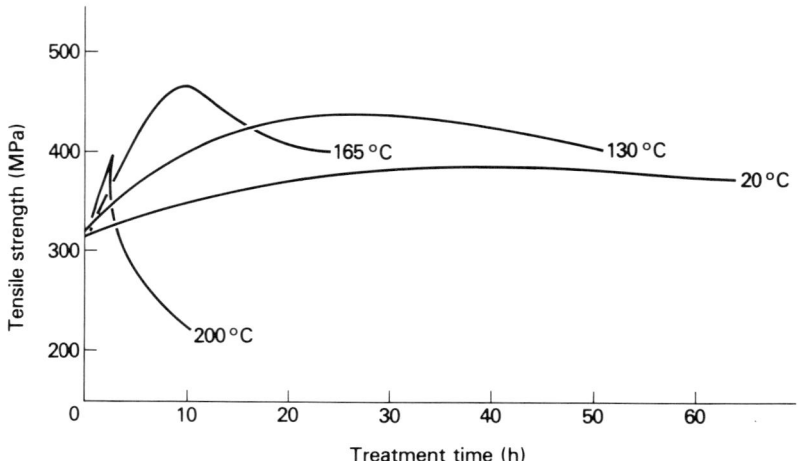

Fig. 7.4 Effects of time and temperature on the precipitation hardening of aluminium alloys

Table 7.6 Heat treatable, cast aluminium alloys

Type	Composition		Condition	Properties			Applications
	Cu	Other Elements		0.1% P.S. (MPa)	U.T.S. (MPa)	Elong^n (%)	
BS 1490:LM4	3.0	Si 5.0 Mn 0.5 Fe 0.8 Ti 0.2	Solution treated at 520 °C for 6 hours. Precipitation hardened at 170 °C for 12 hours	252	294	1	General purpose alloy for sand casting; gravity and pressure die-casting. Withstands moderate stress, shock and hydraulic pressure.
BS 1490:LM8	(Si) (4.5)	Mg 0.5 Mn 0.5 Ti 0.15	Solution treated at 465 °C for 8 hours. Precipitation hardened at 165 °C for 10 hours.	—	280	2	Good casting properties and corrosion resistance. Mechanical properties can be varied by heat treatment.
BS 1490:LM14	4.0	Si 0.3 Mg 1.5 Ni 2.0 Ti 0.2	Solution treated at 510 °C. Precipitation hardened in boiling water for 2 hours.	215	280	—	Pistons and cylinder heads for liquid and water cooled engines. A good general purpose alloy. The original alloy.
BS 1490:LM16	1.2	Si 5.0 Mn 0.5 Ni 0.25	Solution treated at 520 °C for 12 hours; water quenched. Precipitation hardened at 150 °C for 10 hours.	182	231	1	Suitable for castings of intricate shape. High pressure tightness: suitable for valve bodies and cylinder heads. Also used for water jackets and cylinder blocks.

solution that increases the strength and hardness of the alloy. As stated in section 7.3, the heat-treatable alloys are available as casting alloys and wrought alloys.

Casting alloys

These are usually complex alloys containing copper or nickel or both in significant amounts plus other alloying elements in lesser amounts. A number of these alloys contain up to 4 per cent copper, whilst others contain up to 2 per cent nickel. They are softened and grain refined after casting by solution treatment and hardened and strengthened by precipitation hardening when intermetallic compounds such as $CuAl_2$ and $NiAl_3$ are present. Unfortunately these compounds make the alloys less ductile. Examples of these heat-treatable casting alloys are given in Table 7.6, together with their composition and typical applications.

Wrought alloys

These are complex alloys of aluminium together with such alloying elements as copper, magnesium, silicon and zinc. One of the most popular of the heat-treatable wrought alloys is duralumin. It is strong and tough, yet can be cold-worked and easily machined after solution treatment. A typical composition is given in Table 7.7. The presence of copper in the alloy reduces its corrosion resistance, and a composite material called 'Alclad' is available. This consists of a high-strength alloy core, such as duralumin, clad with high purity aluminium as shown in Figure 7.5.

Table 7.7 gives examples of heat treatable wrought alloys, together with their composition and typical applications.

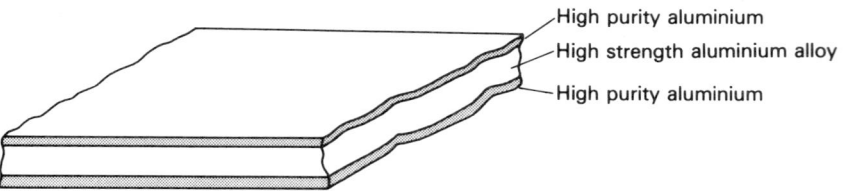

Fig. 7.5 Section of 'Alclad' sheet

7.5 Copper

This is one of the few non-ferrous metals that has sufficient strength to be used as a structural material. Very pure copper has a density of $8.93 \, \text{g mm}^{-3}$. It is very ductile and can be readily drawn into fine wires and solid drawn tubes. It has excellent electrical and thermal conductivity, and it has good corrosion resistance. Like aluminium, it reacts

Table 7.7 Heat-treatable, wrought aluminium alloys

Type	Composition		Condition	Properties			Applications
	Cu	Other Elements		0.1% P.S. (MPa)	U.T.S. (MPa)	Elong^n (%)	
DTD 372	—	Si 0.5 Mg 0.6 Ti 0.2	Solution treated at 520 °C; quenched. Precipitation hardened at 170 °C for 10 hours.	168	224	18	Good corrosion resistance. Extruded sections such as glazing bars and window sections. Windscreen and sliding roof sections for automobile body-building industry.
BS 1470/7 H114	4.0	Mg 0.8 Si 0.5 Mn 0.7	Solution treated at 480 °C; quenched. Age hardened at room temperature for 4 days.	280	400	10	General purpose alloy suitable for stressed parts in aircraft and other structures. The original 'Duralumin'.
BS 1470/7:H730	—	Mg 1.0 Si 1.0 Mn 0.7	Solution treated at 510 °C; quenched. Precipitation hardened at 175 °C for 10 hours.	150	250	20	Structural members for road and rail vehicles and ship building. Architectural work. Ladders and scaffold tubes. High electrical conductivity: overhead powerlines.
DTD 5074	1.6	Mg 2.5 Zn 6.2 Ti 0.3	Solution treated at 465 °C; quenched. Precipitation hardened at 120 °C for 24 hours.	590	650	11	Strongest commercial alloy. Highly stressed aircraft components. Military equipment.

with atmospheric oxide to form a thin, homogeneous oxide film on a freshly cut surface and this prevents further oxidation. High conductivity copper produced by electrolytic refinements has a purity better than 99.9 per cent. This is called cathode copper because the refined copper forms the cathode of an electrolytic cell as shown in Fig. 7.6. The impure copper forms the anode and the electrolyte is a solution of copper sulphate. On the passage of a direct current, copper from the copper sulphate solution is deposited on the cathode. This upsets the chemical balance of the solution, which then dissolves copper from the anode to restore its balance. The process continues as long as the current flows, with very pure copper being deposited on the cathode, and the anode being gradually eaten away. Any impurities are precipitated on the bottom of the electrolytic cell as a dross which periodically has to be flushed away.

For structural purposes, pure copper is insufficiently strong and impurities in the form of copper oxides are allowed to form in the copper. Such coppers are referred to as *tough pitch* and are used for general purpose sheets, rods, tubes, etc. Tough pitch copper does not have such a good electrical conductivity or corrosion resistance as cathode (high purity) copper.

Pure copper is a difficult material to machine to a good surface finish. However the addition of traces of tellurium or sulphur produces a *free-cutting copper* with only slightly impaired ductility and conductivity. For example, the addition of 0.5 per cent tellurium forms the chip-breaking compound copper telluride whilst maintaining an electrical conductivity of 90 per cent copper. Similarly 0.4 per cent sulphur produces the chip breaking compound copper sulphide, whilst retaining an electrical conductivity of 95 per cent. However, the use of sulphur reduces the chip-breaking compound copper sulphide, whilst retaining an To prevent gassing and porosity during welding, phosphorus is added in small quantities to the copper. The phosphorus combines with any dissolved oxygen present, and copper treated in this manner is said to

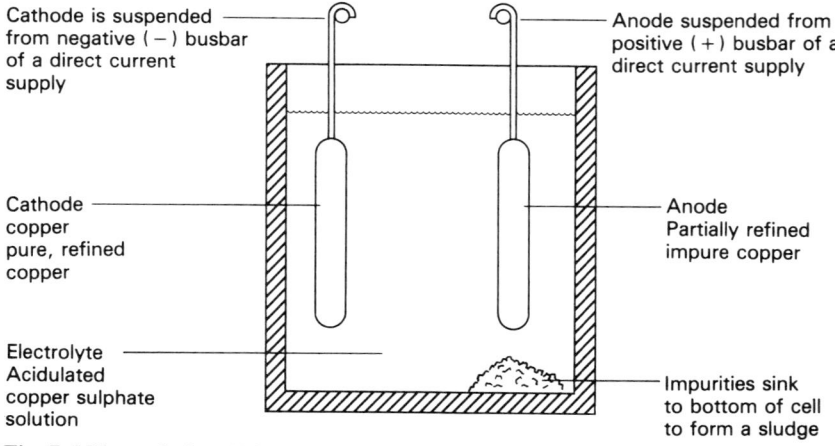

Fig. 7.6 Electrolytic cell for the refinement of copper

be phosphorus deoxidised. Table 7.8 shows the relationship and some typical uses of the more commonly available forms of copper, whilst Table 7.9 gives some typical properties.

Table 7.8 Copper

Copper

↓

Properties:
1. Relatively high strength.
2. Very ductile (easily cold-worked).
3. Corrosion resistant.
4. Second only to silver as a conductor of heat and electricity.
5. Easily joined by soldering, brazing and welding.

↓

One of the few metals of use to the engineer as a structural material in the pure state, although commercial grades contain some trace elements.
Availability: Cold drawn rods, bars, wire and tubes. Cold rolled sheet and strip. Extruded sections. Castings. Powder for sintered components.

↓

Cathode copper
Used in the production of copper alloys.

↓

High conductivity copper
Better than 99.9% pure. Used for electrical conductors and heat exchangers.

↓

Refined tough pitch copper
General purpose copper. Used for roofing, chemical plant, general presswork, decorative metalwork and applications where special properties are not required.

Phosphorus-de-oxidised non-arsenical copper
Welding quality copper. Removal of the dissolved oxygen content prevents gassing and porosity. Used for fabrication, casting, cold impact extrusion and severe presswork.

↓

Arsenical tough pitch and phosphorus de-oxidised copper
The addition of arsenic improves the strength at high temperatures. Used for boiler and firebox plates, stays, flue tubes and domestic plumbing.

Table 7.9 Properties of copper

Description	Purity	Oxygen Content	Condition	U.T.S. (MPa)	Elong^n (%)	Hardness (H_B)
Electrolytic tough pitch high-conductivity copper	99.90% (min)	0.05%	Annealed	220	50	45
			Hard	400	4	115
Fire refined tough pitch high-conductivity copper	99.85% (min)	0.05%	Annealed	220	50	45
			Hard	400	4	115
Oxygen-free high-conductivity copper	99.05% (min)	—	Annealed	215	60	45
			Hard	340	6	115
Phosphorus de-oxidised copper	99.85% (min)	O_2 nil P 0.013 to 0.05%	Annealed	210	60	145
			Hard	320	4	115
Arsenical copper	99.20% (min)	0.05% O_2 0.3 to 0.5 As	Annealed	220	50	45
			Hard	400	4	115

7.6 High copper content alloys

The first group of copper alloys to be considered are the high copper content alloys. That is, the additional alloying element represents only a very small percentage of the total; yet these small additions make a significant change to the properties of the alloy compared with pure copper.

Silver copper

The addition of only 0.1 per cent of silver to high conductivity copper raises its annealing temperature by up to 150 °C. This is very important where electrical conductors have to be soldered to hard-drawn copper contacts. If pure copper is used the heat required to make the soldered joint would soften the copper and render it useless as a contact material.

Cadmium copper

Like silver, cadmium has little effect upon the conductivity of copper. Again like silver, it also raises the softening temperature. In addition, however, it also strengthens and toughens the copper, increasing its resistance to fatigue. As cadmium copper is substantially free from oxygen it is not susceptible to 'gassing' when it is braze welded.

Cadmium copper is used for low and medium voltage overhead transmission lines where its high conductivity and high tensile strength enable it to be used over relatively long spans.

Cadmium copper is also used for traction purposes, e.g. the overhead conductors on the electrified lines of railways. It is also recommended in the soft condition, for aircraft wiring where its flexibility is combined with its resistance to the effects of vibration.

Chromium copper

This is one of the few non-ferrous materials that can be heat treated to improve its mechanical properties rather than relying upon cold work-hardening. A typical alloy contains 0.5 per cent chromium and is quenched from 100 °C. This leaves the alloy in a soft and ductile condition with rather a low electrical conductivity. However, if the metal is reheated to 500 °C for approximately two hours, the mechanical and electrical properties are improved.

Since the properties of chromium copper depend upon heat treatment rather than cold-working, it can be used in cast as well as wrought forms. Similarly, components can be formed from annealed sheets or extruded rod and then hardened after manipulation and machining.

Tellurium copper

Tellurium forms stable compounds with copper and the addition of 0.5 per cent makes the copper as machineable as a free-cutting brass whilst retaining a high electrical conductivity. It has a very high corrosion resistance and is used extensively for heavy-duty contacts and commutators on machines and switchgear situated in hostile environments such as mines and chemical plants. The addition of traces of nickel and silicon allows it to be heat-treated similarly to chromium copper.

Beryllium copper

Beryllium copper is used where mechanical rather than electrical properties are required. Beryllium copper is softened by heating to 800 °C and quenching. In this condition the material can be extensively cold-worked. Subsequent heat treatment consists of heating the metal to 300 to 320 °C for upwards of two hours. The resulting mechanical properties will depend to some extent upon the degree of cold-working that took place between the first and second treatments.

Beryllium copper is used widely for instrument springs, flexible bellows, corrugated diaphragms (aneroid altimeters and barometers) and the bourdon tubes for pressure gauges.

Because hand tools made from beryllium copper are almost as strong as those made from steel, but will not strike sparks from other metals or from flint stones, such tools are widely used in hazardous locations where there is a high risk of explosion such as mines, oil refineries, oil rigs, chemical plant, etc. Its high cost precludes its use for more conventional situations.

7.7 Copper alloys

Unlike the alloys discussed in section 7.6, the alloys to be considered in this section contain elements present in much greater amounts. For example, in some brasses the zinc content may make up to 40 per cent of the alloy. Four groups of alloys will be considered in this section, namely:

(a) the brass alloys (copper-zinc);
(b) the tin-bronze alloys;
(c) the aluminium-bronze alloys;
(d) the cupro-nickel alloys.

The brass alloys

These are alloys of copper and zinc. They tend to give poor quality porous castings which depend upon hot or cold working to consolidate the metal and improve its mechanical properties. The more common brasses are listed in Table 7.10.

The change in properties with change in composition are more readily understood by reference to the copper-zinc thermal equilibrium diagram shown in Fig. 7.7. Although more complex than any thermal equilibrium diagram considered so far, it is fairly easy to understand if taken section by section. As in previous diagrams the α phase consists of a solid solution, which in this case is a solid solution of zinc in copper. Like all solid solutions it is ductile and suitable for cold-working.

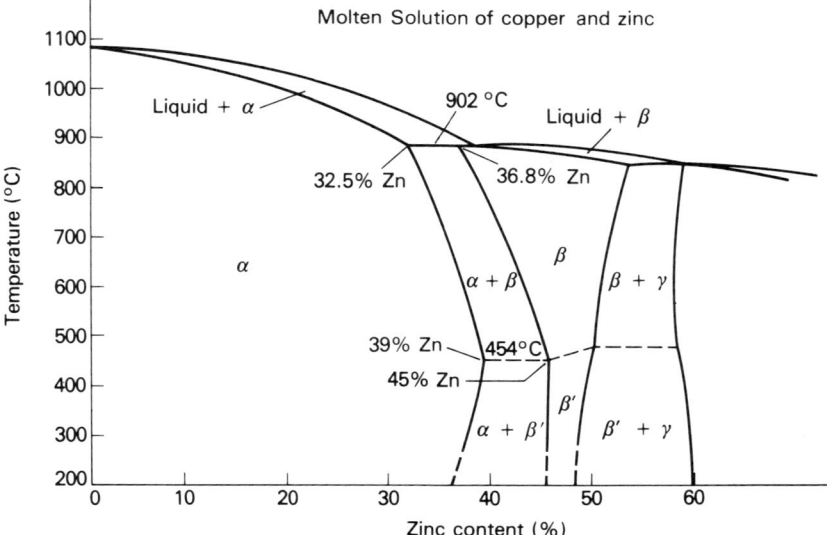

Fig. 7.7 Copper-zinc thermal equilibrium diagram

Table 7.10 Brass alloys

Name	Composition (%)			Applications
	Copper	Zinc	Other elements	
Cartridge brass	70	30	—	Most ductile of the copper–zinc alloys. Widely used in sheet metal pressing for severe deep drawing operations. Originally developed for making cartridge cases, hence its name
Standard brass	65	35	—	Cheaper than cartridge brass and rather less ductile. Suitable for most engineering processes
Basis brass	63	37	—	The cheapest of the cold working brasses. It lacks ductility and is only capable of withstanding simple forming operations
Muntz metal	60	40	—	Not suitable for cold-working but hot-works well. Relatively cheap due to its high zinc content, it is widely used for extrusion and hot-stamping processes
Free-cutting brass	58	39	3% lead	Not suitable for cold-working but excellent for hot-working and high-speed machining of low strength components
Admiralty brass	70	29	1% tin	This is virtually cartridge brass plus a little tin to prevent corrosion in the presence of salt water
Naval brass	62	37	1% tin	This is virtually Muntz metal plus a little tin to prevent corrosion in the presence of salt water

When the amount of zinc present exceeds that required to saturate the α phase solid solution, a β phase is introduced. This new phase is tougher, stronger and harder than the α phase but also less ductile. The β' phase can only exist down to 453 °C below which it is modified to β' phase. This lowers the ductility and malleability still further. If even more zinc is added an even more brittle and less ductile phase is introduced called γ brass. Commercial brasses usually contain only α or α + β phases. The effect of these phases on the properties of a range of brasses are shown in Fig. 7.8, and these properties should be compared with the applications listed in Table 7.10.

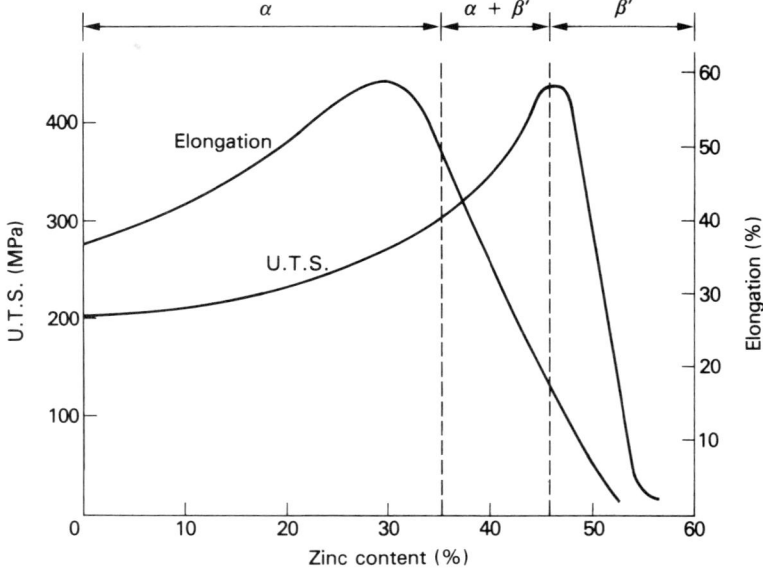

Fig. 7.8 Effect of composition on the properties of brass

The α + β brasses are largely used for hot-working such as hot stamping and extrusion. This is because the β phase is much more ductile and malleable than the β' phase. The addition of lead or tellurium to (duplex) brasses gives them free-cutting properties so that they machine easily. These additional alloying elements have little effect on the strength, hardness and ductility of the brass.

The phases found in the copper-zinc alloy system may be summarised as follows:

α phase → Zinc (Zn) in excess copper (Cu)
β phase → Copper (Cu) in excess zinc (Zn)
(strictly a brittle intermetallic compound (CuZn))
γ phase → A complex cubic structure which produces extreme embrittlement in the alloy and is rarely found in commercial brasses

Table 7.11 Tin-bronze alloys

Name	Composition (%)				Application
	Copper	Zinc	Phosphorus	Tin	
Low-tin bronze	96	—	0.1 to 0.25	3.9 to 3.75	This alloy can be severely cold-worked to harden it so that it can be used for springs where good elastic properties must be combined with corrosion resistance, fatigues resistance and electrical conductivity, e.g. contact blades
Drawn-phosphor-bronze	94	—	0.1 to 0.5	5.9 to 5.5	This alloy is used in the work-hardened condition for turned components requiring strength and corrosion resistance, such as valve spindles
Cast phosphor-bronze	Rem.	—	0.03 to 0.25	10	Usually cast into rods and tubes for making bearing bushes and worm wheels. It has excellent anti-friction properties
Admiralty gunmetal	88	2	—	10	This alloy is suitable for sand casting where fine-grained, pressure-tight components such as pump and valve bodies are required
Leaded-gunmetal (free-cutting)	85	5 (5% lead)	—	5	Also known as 'red brass', this alloy is used for the same purposes as standard, admiralty gunmetal. It is rather less strong but has improved pressure tightness and machining properties
Leaded (plastic) bronze	74	(24% lead)	—	2	This alloy is used for lightly loaded bearings where alignment is difficult. Due to its softness, bearings made from this alloy 'bed in' easily

Tin-bronze alloys

These are alloys of copper and tin together with a deoxidiser. The deoxidiser is essential to prevent the tin content oxidising during casting and hot-working. Oxidation of the tin would result in a weak, brittle, 'scratchy' bronze. Two deoxidisers commonly used are:

(a) phosphorus in the 'phosphor-bronze' alloys;
(b) zinc in the 'gunmetal' alloys.

Some typical tin-bronze alloys are listed in Table 7.11.

Unlike the brasses which are largely used in the wrought condition (rod, sheet, etc.), only low tin content bronzes can be worked and most bronze components are in the form of castings. Tin bronzes are more expensive than the brasses, but are stronger and give sound, pressure-tight castings that are widely used for steam and hydraulic valve bodies and mechanisms. They are highly resistant to wear and corrosion.

The thermal equilibrium diagram for copper-tin alloys is shown in Fig. 7.9. Again it is a complicated diagram; yet again it is not too difficult to understand if taken section by section. Up to about 10 per cent tin, the composition is largely α phase solid solution of tin in copper (although it is obvious that this changes with temperature). Thus these low tin content alloys are ductile and can be cold-worked.

Care must be taken when making bronze alloys to prevent oxidation as this will weaken the alloy and make it 'scratchy' and brittle.

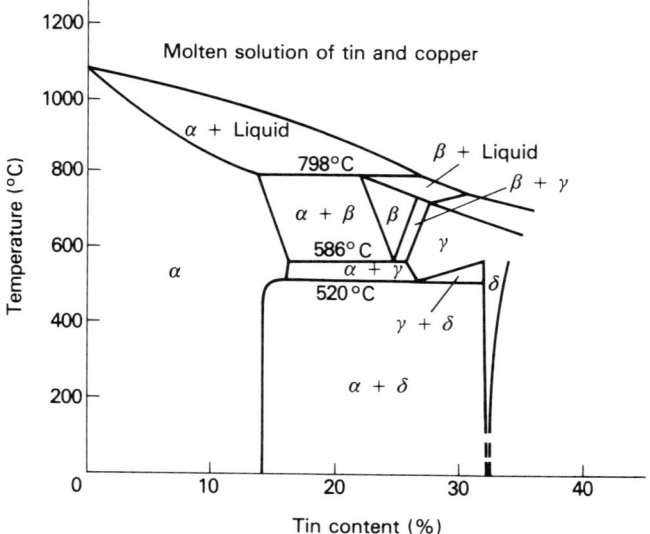

Fig. 7.9 Copper-tin thermal equilibrium diagram (copper-rich alloys)

Therefore in all tin bronzes a deoxidising agent is introduced. In the low tin bronzes this is always phosphorus and the resulting alloy is called a *phosphor bronze*. After work hardening, by cold rolling or cold drawing, the resulting strip or wire is used for instrument springs, electrical contacts, and similar components.

If the tin content is increased, the γ phase is introduced and also a copper-phosphorus compound. The γ phase is a hard brittle intermetallic compound of copper and tin. This makes the alloy unsuitable for hot- or cold-working, but gives it excellent casting properties. The combined hardness and low friction properties of the cast phosphor-bronze alloys makes them ideal for bearing bushes, high grade gears and similar components.

Alternatively, zinc may be used instead of phosphorus as a deoxidiser. Casting bronzes containing zinc are called Admiralty gunmetals. These alloys give excellent pressure-tight castings with a high corrosion resistance. They are widely used for valve bodies and marine castings – hence their name.

The more important phases found in the copper-tin alloy system may be summarised as follows:

α phase→tin (Sn) in excess copper (Cu)
β phase→copper (Cu) in excess tin (Sn)
δ phase→a hard brittle intermetallic compound $Cu_{31}Sn_8$

Aluminium-bronze alloys

These are more expensive than 'tin bronzes' but are more corrosion resistant at high temperatures. They are also more ductile and can be cold-worked into tubes for boilers and condensers in steam plant.

Figure 7.10 shows the aluminium-bronze thermal equilibrium diagram. Once again it is the α phase solid solution of aluminium in copper which produces the ductile, cold-working alloys which can be rolled into sheet and drawn into tubes. Over about 10 per cent, duplex alloys consisting of $\alpha + \beta$ phases above 550 °C and $\alpha + \gamma_2$ below 550 °C are produced. The γ_2 phase is a hard brittle intermetallic compound Cu_9Al_4. Thus alloys containing this compound are unsuitable for cold working but are excellent for casting.

The aluminium-bronze thermal equilibrium diagram is particularly interesting since it is analogous to the iron-carbon equilibrium diagram. The α phases in both diagrams are directly analogous, and the β phase of Fig. 7.10 corresponds to the γ_2 phase (austenite) of the iron-carbon diagram. Similarly, the $\alpha + \gamma_2$ eutectoid corresponds to pearlite (α + cementite) in steels. Because of these similarities 10 per cent aluminium bronze alloys can be heat-treated like steel.

For example, cooling in a 10 per cent alloy under equilibrium conditions (annealing) results in the formation of $\alpha + \gamma_2$ at room temperature. On heating above 565 °C the $\alpha + \gamma_2$ eutectoid transforms into the

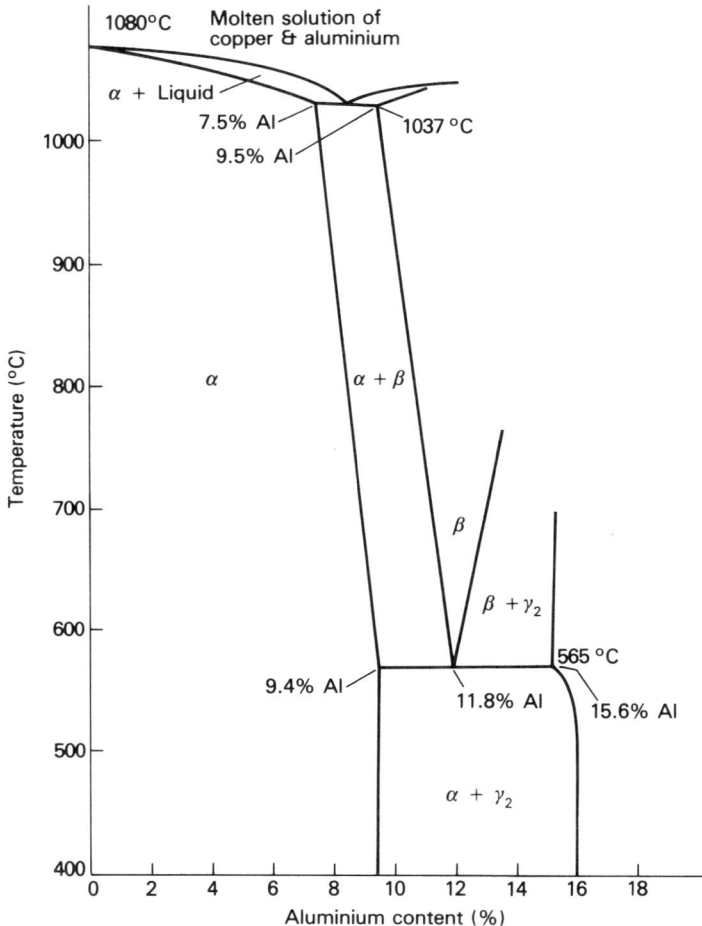

Fig. 7.10 Copper-aluminium thermal equilibrium diagram (copper-rich alloys)

β phase solid solution and on further heating to 900 °C the remaining α phase is transferred into β phase. Quenching in water from this temperature produces a structure of the β phase. This is not shown in Fig. 7.10 since it is not an equilibrium transformation. The β phase is analogous to martensite being very hard and brittle, it also has the same acidular (needle-like) appearance under the microscope. The β' phase can be tempered at 500 °C when a fine agglomeration of α and γ_2 is precipitated, which is analogous to sorbite in tempered steels.

Aluminium bronzes are highly corrosion resistant, particularly at high temperatures, because of the film of homogeneous aluminium oxide which forms on the metal surface, and which is self-healing if damaged. Table 7.12 lists some examples of aluminium bronze alloys together with their composition and typical applications.

Cupro-nickel alloys

Copper and nickel form alloys in which the two metals are wholly soluble in each other both in the liquid and the solid state. Because of this, their thermal equilibrium diagram was considered in depth in Chapter 2. Since only the α phase is present over the entire range of alloys, they all have high strength and ductility and are suitable for hot- and cold-working. These are relatively expensive copper-based alloys, but their strength and resistance to corrosion makes them highly suitable for such applications as high duty boiler and condenser tubes, bullet envelopes and resistance wires.

Monel metal, which also contains traces of iron and manganese, has exceptionally high corrosion resistance and is used for chemical plant and for marine applications.

Nickel silvers contain up to 20 per cent zinc and have a silvery appearance. These alloys can be readily cold-worked and are used for making knives, forks and spoons. Table 7.13 lists some typical cupro-nickel alloys together with their composition and typical applications.

7.8 Magnesium alloys

Magnesium is the lightest of the engineering metals with a density of only 1.7 g mm^{-3}. Its electrical conductivity is 60 per cent that of high conductivity copper and it has a high thermal conductivity. It has a high affinity for atmospheric oxygen and magnesium foil burns in air with a fierce white flame. It was widely used at one time for flash photography. Because of this, and because of its low tensile strength, magnesium is only used for engineering applications as the basis of a range of ultralight alloys.

Although not as strong as aluminium alloys, magnesium alloys have a much lower density, so that in many instances their strength-to-weight ratio is superior. The alloys fall into two main categories:

1. casting alloys;
2. wrought alloys.

Magnesium alloys contain aluminium, zinc, zirconium and manganese, together with 'rare-earth' metals in some instances. Like the aluminium alloys, magnesium alloys can be annealed (softened) by solution treatment at 380 °C for eight hours, followed by sixteen hours at 410 °C. They can also be precipitation hardened at 190 °C for some ten to twelve hours. The melting and casting of magnesium alloys is difficult in view of the flammability of magnesium if accidentally overheated. They are generally melted under a flux containing calcium and sodium fluorides to exclude atmospheric oxygen. As the molten metal is poured it is dusted with flour of sulphur. The sulphur burns and blankets the molten alloy in sulphur dioxide gas which excludes atmospheric oxygen without which the molten alloy cannot ignite. Table 7.14 lists some typical cast and wrought magnesium alloys.

Table 7.12 Copper–aluminium (aluminium bronze) alloys

Composition (%)				Condition	Properties				Applications
Al	Fe	Cu	Other Elements		0.1% P.S. (MPa)	U.T.S. (MPa)	Elong^n (%)	Hardness (H_D)	
5	—	Rem.	Mn or Ni Up to 4%	Annealed	112	350	70	80	Cold-worked for decorative purposes such as imitation jewellery. Tubes for engineering applications, resistant to corrosion and oxidation.
				Hard	532	700	4	200	
8	—	Rem.	Fe, Mn, Ni Up to 2%	Hot-worked	140	392	45	100	Chemical engineering plant suitable for use at moderately elevated temperatures.
10	5	80	Ni–5%	Hot-forged	420	658	20	215	General engineering forgings combining strength and corrosion resistance – can be heat treated.
9.5	2.5	Rem.	Ni and Mn Up to 1.0%	As cast	168	476	30	115	General purpose alloy for both die-casting and sand-casting.
12	—	Rem.	Fe, Mn, Ni from 5% to 8%	As cast	434	504	3	250	A hard, rigid alloy containing the γ_2 phase. Used where heavy compressive loads are involved. Good wear resistance

Table 7.13 Cupro-nickel alloys

Composition (%)			Condition	Properties				Applications
Cu	Ni	Other Elements		0.1% P.S. (MPa)	U.T.S. (MPa)	Elong^n (%)	Hardness (H_D)	
80	20	—	Annealed	98	308	45	75	Very high ductility and corrosion resistance, will withstand severe cold working.
			Hard	420	490	5	165	
70	30	—	Annealed	98	322	45	80	Used for condenser and heat-exchanger tubes where high corrosion resistance is required.
			Hard	490	588	5	175	
60	40	—	Annealed	—	350	45	90	'Constantan' electrical resistance wire – high specific resistance and low temperature coefficient. Also used in thermocouples.
			Hard	—	588	5	190	
29	68	Fe 1.25 Mn 1.25	Annealed	196	490	45	120	'Monel Metal'. Good mechanical properties combined with excellent corrosion resistance properties. Used for chemical plant.
			Hard	518	658	70	220	

Table 7.14 Magnesium alloys

	Composition (%) (Remainder Mg)						Condition	Properties			Applications
	Al	Mn	Zn	Zr	Th	Rare earths		0.1% P.S. (MPa)	U.T.S. (MPa)	Elong^n (%)	
Casting alloys	10.0	0.3	0.7	—	—	—	Chill-cast	115	200	2	Light-weight castings for the aircraft and high performance car industry. For example:– Landing wheels, road wheels, crank cases, and miscellaneous coatings.
	—	—	4.0	0.7	—	1.2	As Cast Heat-treated	95 130	170 215	5 4	
	—	—	—	0.7	3.0	—	Heat-treated	100	210	8	
Wrought alloys	—	1.5	—	—	—	—	Rolled	95	200	5	Petrol tanks, oil tanks and other lightly stressed sheet metal components. Light-weight sections.
	—	1.0	—	—	3.0	—	Rolled	215	280	10	
	6.0	0.3	1.0	—	—	—	Forged Extruded	155 140	280 215	8 8	Light-weight forgings for the aircraft industry such as airscrew blades and undercarriage components.
	—	—	3.0	0.7	—	—	Forged Extruded	170 215	265 310	8 8	

7.9 Zinc alloys

Pure zinc is an interesting metal, its boiling point being so low that it is the only commercial metal that is refined by distillation. It has a density of 7.1 g mm^{-3} and a melting point of only 420 °C. The pure metal is relatively weak, but it is widely used as a coating on steel to prevent corrosion from atmospheric attack. Zinc-coated steel is known as galvanised steel. In the presence of dilute acids (which abound in the atmosphere from the combustion of fossil fuels) it is *sacrificial*, and the zinc is slowly eaten away in preference to the iron. It forms a good key for a paint system and if galvanised iron is properly and regularly painted, the steel so treated should remain corrosion free almost indefinitely.

Zinc-based alloys are used almost entirely for pressure die-casting. They are widely used for car door handles, carburettor and fuel pump bodies, and other lightly stressed complex components. Zinc pressure die-castings can be machined, but they cannot be soldered or welded. The popularity of zinc-based alloys for die-casting is due to their

(*a*) high fluidity which enables complex castings with thin sections to be made;
(*b*) low melting point which reduces die wear;
(*c*) narrow freezing range which allows high rates of production;
(*d*) easy polishing and electroplating properties;
(*e*) adequate strength for small components.

Zinc-based alloys normally contain approximately:

Aluminium	4%
Copper	2.7%
Zinc	remainder

The zinc must be better than 99.99 per cent pure, otherwise even minute traces of metals such as cadmium, tin and lead will lead to intercrystalline brittleness, swelling and corrosion. It was the lack of high purity zinc which gave zinc alloy die-castings their poor reputation for quality in the early days of the process. Zinc die-castings should have a strength of about 320 N mm^{-2} and a high rigidity.

The corrosion resistance of zinc based die-castings can be increased by the 'chromate passivation' process. The castings are cleaned and immersed in a solution of dilute sulphuric acid and sodium bichromate for not more than one minute. They are then washed and dried. The resultant passive film prevents atmospheric corrosion in damp atmospheres for those applications where the die-casting is not electroplated.

The range of alloys discussed in this chapter are only intended as a representation of the range available. Any one category can be a study in its own right. In addition are the titanium alloys used in modern aircraft construction, and the nickel-chromium alloys used for electrical resistance and heating element wires and other corrosion-resistant, high-temperature applications. The soft solders (tin-lead alloys) and the white bearing metals were considered in Chapter 2.

Problems
Section A
1. State the main alloying elements in: (i) brass; (ii) gun metal; (iii) duralumin; (iv) phosphor bronze.
2. Describe the effect on the ductility and strength of aluminium of: (i) the purity of the aluminium; (ii) the amount of cold work it has received.
3. With reference to aluminium alloys state the main alloying elements to be found in: (i) heat-treatable casting alloys; (ii) heat-treatable wrought alloys; (iii) non-heat-treatable casting alloys; (iv) non-heat-treatable wrought alloys.
4. Select a suitable non-ferrous metal or alloy for each of the following applications giving reasons for your choice: (i) an electrical conductor which can be soldered; (ii) a die-cast car door handle; (iii) an internal combustion engine piston; (iv) a heavy-duty bearing bush.
5. State the effect on copper of adding the following alloying elements in small quantities: (i) tellurium; (ii) beryllium; (iii) cadmium; (iv) arsenic.

Section B
6. (a) Sketch the aluminium-silicon thermal equilibrium diagram and explain in detail the changes that take place as a 6 per cent silicon alloy is cooled from the molten state of room temperature.
   (b) Discuss the need for the 'modification' of aluminium-silicon alloys by the addition of metallic sodium, and show the effect of modification on the aluminium-silicon thermal equilibrium diagram.
7. Discuss in detail the differences in composition, properties and applications of: (i) high conductivity copper; (ii) fire-refined tough pitch copper; (iii) oxygen-free high-conductivity copper; (iv) phosphorus deoxidised copper.
8. (a) State the composition, properties and typical applications of the aluminium alloy known as 'duralumin'.
   (b) Describe in detail, with reference to the aluminium copper equilibrium diagram, the softening of duralumin by solution treatment, and the subsequent hardening of duralumin by natural (precipitation) ageing.
9. Sketch the copper-zinc thermal equilibrium diagram and refer to it when describing the differences in composition, properties and typical applications of: (i) the α brasses; (ii) the duplex brasses $(\alpha + \beta')$.
10. Sketch the copper-tin thermal equilibrium diagram and refer to it when describing the differences in composition and typical applications of: (i) the α phase bronze alloys; (ii) the $\alpha + \delta$ duplex bronze alloys.

Chapter 8

Polymers

8.1 Polymeric materials

There is an ever-increasing number of synthetic, polymeric materials available under the name of *plastics*. This name is a misnomer since polymeric materials rarely show plastic properties in their finished condition, in fact many of them are elastic, but during the moulding operation by which they are formed they are reduced to a plastic condition by heating to just above the temperature of boiling water.

There are two main groups of plastic materials:

Thermosetting plastics

These undergo a chemical change during moulding and can never be softened by reheating. These materials are hard, rigid and rather brittle. Some typical examples are given in Table 8.1. The strength of thermosetting plastics can be greatly increased by reinforcing them with fibrous materials.

Thermoplastics

These can be softened as often as they are reheated. They are not so rigid as the first group and tend to be tougher. Some typical examples are given in Table 8.2.

Laminated plastic (Tufnol)

Fibrous material such as paper, woven cotton cloth or woven asbestos cloth is impregnated with phenolic resin (this is the basic material of bakelite). The sheets of fabric are then laid up in hydraulic presses and squeezed and heated so that they become solid sheets, rods, tubes, etc.

Table 8.1 Some typical thermosetting plastic materials

Material	Characteristics
Phenolic resins and powders	These are used for dark-coloured parts because the basic resin tends to become discoloured. These are heat-curing materials
Amino (containing nitrogen) resins and powders	These are colourless and can be coloured if required; they can be strengthened by using paper-pulp fillers, and used in thin sections
Polyester	Polyester chains can be cross-linked by using a monomer such as styrene; these resins are used in the production of glass-fibre laminates
Epoxy resins	These are also used in the production of glass-fibre laminates

This material can be machined dry with ordinary engineering machine tools, using low rake tools and fairly high cutting speeds. Care must be taken because of the 'grain' of the material which causes it to behave rather like plywood.

It is widely used for making bearings, gears and other engineering components.

Glass-reinforced plastic (GRP)

Woven glass fibre can be bonded together by epoxy and polyester resins to form large and complex mouldings (from crash helmets to 18-metre yachts). The resin used is a thermosetting plastic and it is set by chemical action at room temperature; a press is not required. The impregnated glass fibre is laid up over wooden or plaster patterns. When set, it is lifted off and the pattern can be used again.

This material is used to produce large casting patterns, copymilling models, machine casings and guards.

8.2 Plastic building blocks

All the plastic materials introduced in section 8.1 are built from carbon atoms in association with other elements such as oxygen, hydrogen, nitrogen, chlorine and fluorine. Carbon atoms have four chemical bonds or, as a chemist would say, a valency of four. Hydrogen atoms have a valency of one, so if hydrogen and carbon are combined in the simplest way to give a molecule of methane (natural) gas, the molecule would appear as:

$$\begin{array}{c} H \\ | \\ H-C-H \\ | \\ H \end{array}$$

Table 8.2 Some typical thermoplastic materials

Type	Material	Characteristics
Cellulose plastics	Nitrocellulose	Materials of the 'celluloid' type are tough and water resistant. They are available in all forms except moulding powders. They cannot be moulded because of their inflammability
	Cellulose acetate	This is much less inflammable than the above. It is used for tool handles and electrical goods
Vinyl plastics	Polythene	This is a simple material that is weak, easy to mould, and has good electrical properties. It is used for insulation and for packaging
	Polypropylene	This is rather more complicated than polythene and has better strength
	Polystyrene	Polystyrene is cheap, and can be easily moulded. It has a good strength but it is rigid and brittle and crazes and yellows with age
	Polyvinyl chloride (PVC)	This is tough, rubbery, and practically non-inflammable. It is cheap and can be easily manipulated: it has good electrical properties
Acrylics (made from an acrylic acid)	Polymethyl methacylate	Materials of the 'perspex' type have excellent light transmission, are tough and non-splintering, and can be easily bent and shaped
Polyamides (short carbon chains that are connected by amide groups-NHCO)	Nylon	This is used as a fibre or as a wax-like moulding material. It is fluid at moulding temperature, tough, and has a low coefficient of friction
Fluorine plastics	Polytetrafluoro-ethylene (ptfe)	Is a wax-like moulding material; it has an extremely low coefficient of friction. It is very expensive
Polyesters (when an alcohol combines with an acid, an 'ester' is produced)	Polyethylene terephthalate	This is available as a film or as 'Terylene'. The film is an excellent electrical insulator

Thus four hydrogen atoms combine with one carbon atom to make one molecule of methane gas. This molecule is given the formula CH_4 and, because it consists solely of hydrogen and carbon, it is referred to as a 'hydrocarbon'. The hydrocarbons are found in crude oil, coal, and natural gas and they can be classified into four groups:

1. paraffins;
2. olefins;
3. napthenes;
4. aromatics.

Formula		Name	Use
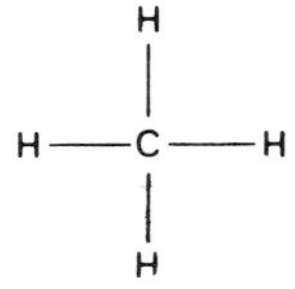 (structure of methane)	CH_4	Methane	Natural gas
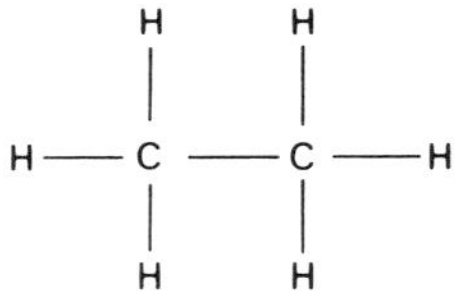 (structure of ethane)	C_2H_6	Ethane	Converted into plastics
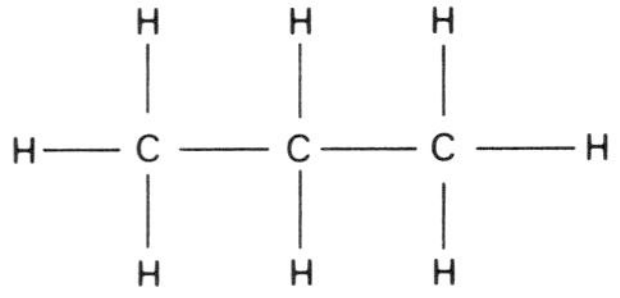 (structure of propane)	C_3H_8	Propane	Heating fuel
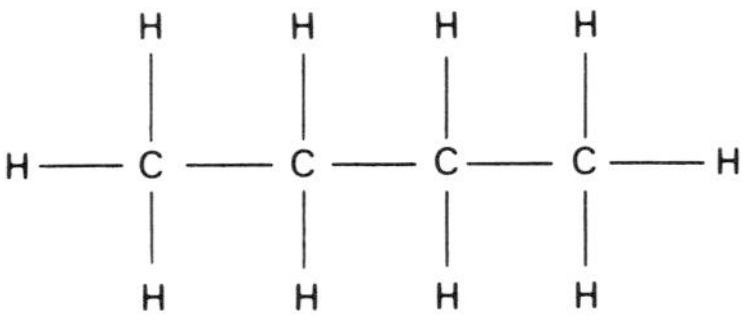 (structure of butane)	C_4H_{10}	Butane	1. Heating fuel 2. Converted into Synthetic rubbers

The series continues to C_{100} to become the asphalts and tars used for roads and roofing

H = Hydrogen.   C = Carbon.

Fig. 8.1 Common paraffins

The paraffins

These are the simplest of the four groups. They have a general formula of C_nH_{2n+2}. For example in the methane molecule just considered there is only one carbon atom so $n=1$, and the number of hydrogen atoms is $2(1)+2=4$. This agrees with the formula already stated as CH_4. Other common paraffins are given in Fig. 8.1. One way of recognising the paraffins is the fact that their names always end in -ane (as in methane, propane, octane, etc.). The paraffins are *saturated* hydrocarbons, that is, they contain the maximum number of hydrogen atoms in each case, as shown in Fig. 8.1, and this makes them rather inactive chemically. The paraffins are the most common group of hydrocarbons appearing in crude oil.

The olefins

These are unsaturated hydrocarbons, that is, additional hydrogen atoms have to be added to olefins to saturate them. This unsaturated condition makes them chemically reactive and olefins form the basis of many plastic and synthetic rubber (elastic) materials. If their general formula is C_nH_{2n}, they are called mono-olefins and are given names ending in -ene or -ylene (as in ethylene, propylene, etc.). There are more complex forms of the olefins, but they need not be considered here. Olefins do not occur often in crude oils and they are usually produced in the course of oil refinery operations and then used as a feed stock for the plastics industry where they are known as *chemical intermediates*. Some typical mono-olefins are shown in Fig. 8.2.

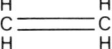

Ethylene (derived from ethane)

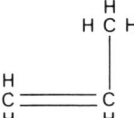

Propylene (derived from propane)

H = Hydrogen
C = Carbon

Fig. 8.2 Common olefins

The naphthenes and aromatics

These both have ring-shaped molecules, as shown in Fig. 8.3. Plastic materials made from a ring-shaped molecule have improved mechanical properties (e.g. the high tensile strength of nylon).

Naphthenes have a saturated molecule with the general formula C_nH_{2n} and names beginning with 'cyclo-' (as in cyclohexane). *Aromatics*, on the other hand, are unsaturated and are chemically highly reactive, being used in solvents and explosives. Aromatics are rare in crude oil (except those found in California), but occur in coal. They form the basis of the styrene group of plastics.

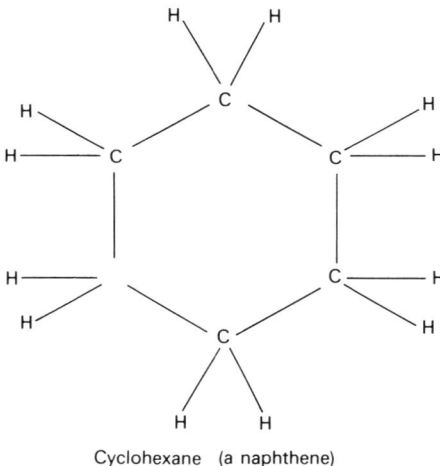

Cyclohexane (a naphthene)

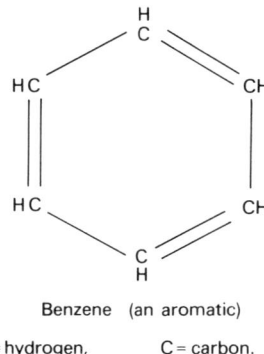

Benzene (an aromatic)

H = hydrogen,   C = carbon.

Fig. 8.3 Common naphthenes and aromatics

To manufacture a plastic material, a paraffin has first to be converted in its corresponding olefin. Hence the term 'chemical intermediate'. A single molecule of the olefin is referred to as *monomer* and the next stage in the process is to combine several monomers to form a much larger molecule called a polymer (*poly* means *many*). In the form of a polymer the olefin takes on the characteristics of a plastic material. Some examples are shown in Fig. 8.4. There are some simple basic rules governing the numbers of monomers that may be brought together to form a polymer. For example, at room temperatures ethylene, which is made up of single molecules (monomers), is a gas. A polymer of six monomers is a liquid; a polymer of 36 monomers is a grease, a polymer of 140 monomers is a wax, and a polymer of 500 or more monomers is the plastic material known as polyethylene. The maximum number of monomers brought together in a single polymer is normally limited to 2000. At his point there is little further increase in strength, but a considerable increase in hardness and brittleness. This rule applies to most plastic materials.

The plastic materials containing only carbon and hydrogen are flammable. To render these materials non-flammable (and the rules governing building applications insist on this) at least one of the hydrogen atoms in each monomer has to be replaced by chlorine, as shown in Fig. 8.5. The resulting polyvinyl chloride is a stiff non-flammable plastic suitable for extruding into rain guttering. Fluorine may be added instead of chlorine to produce a more expensive material with superior mechanical and fire-resistant properties. It is also more resistant to sunlight.

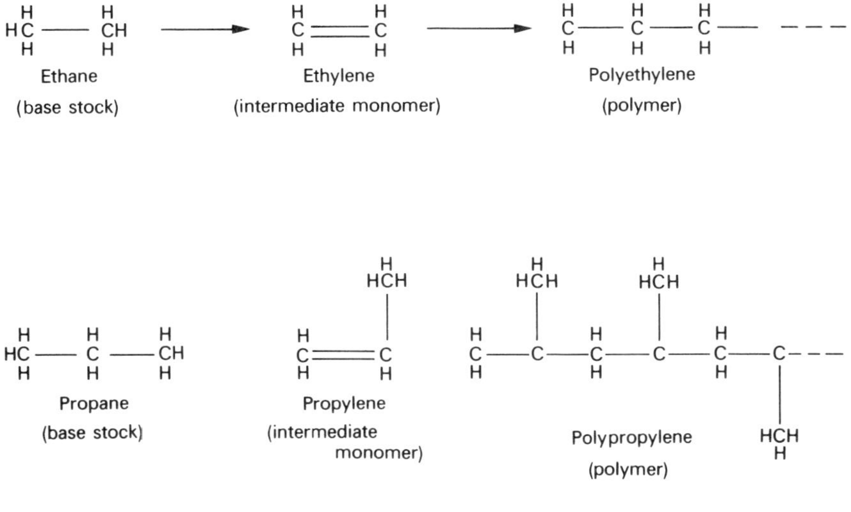

Fig. 8.4 Simple polymers

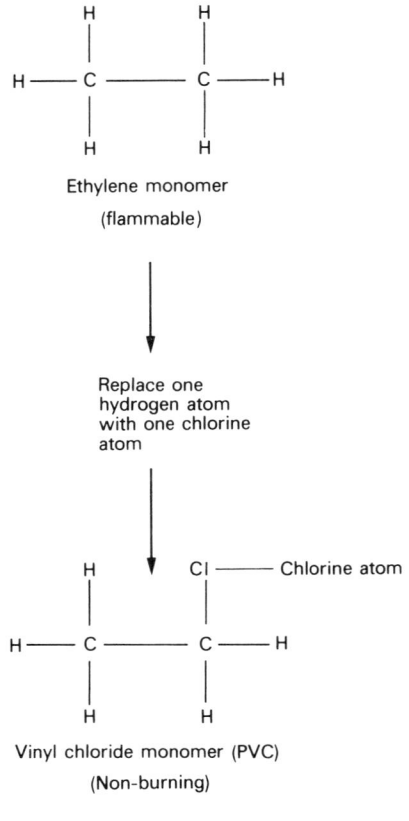

Fig. 8.5 Chlorinated plastics

8.3 Polymers

From section 8.2 it becomes obvious that all plastic materials have two things in common.

(a) They are all made up of long chains of individual unit molecules. These individual unit molecules are called *monomers* and when a large number of these monomers are repeated over and over again in a long chain they are referred to as polymers. Hence such materials are known as *polymeric* materials. This is a much more accurate description of these materials than the popular word *plastic*.

(b) They are all based on a *chain* of carbon atoms which builds up a giant molecule. It is the *shape* of this chain as well as its composition which determines the properties of polymeric materials.

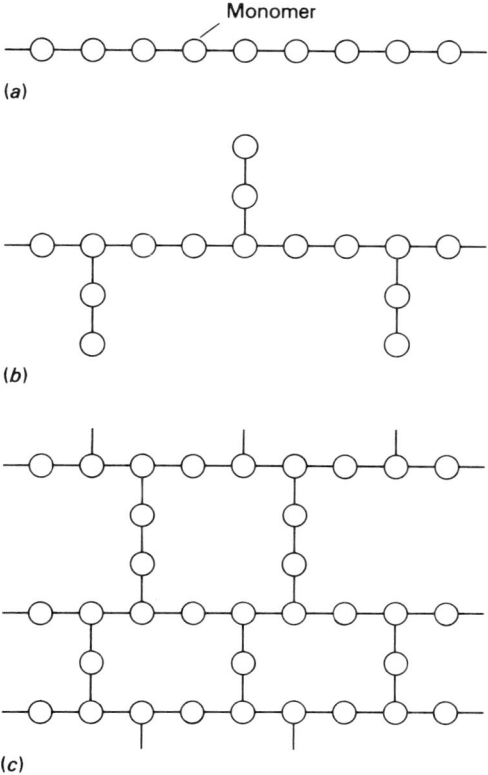

Fig. 8.6 Typical polymer chains (*a*) Linear polymer chain (*b*) Branched polymer chain (*c*) Cross-linked polymer chain

Figure 8.6 shows some of the forms which the molecular chain of a polymeric material may take. The linear chain shown in Fig. 8.6 (*a*) and the linear chain with side branches shown in Fig. 8.6 (*b*) are typical of thermoplastic materials.

The simple linear chains with no side branches can move easily past each other. This results in a non-rigid thermoplastic material which can be easily flexed and stretched. *Polyethylene* is an example of such a material. Since no additional heat energy is required to break down the side branches, materials such as polyethylene melt at low temperatures and easily return to their original state when they cool down.

Since it is more difficult for branched linear chains to move past each other, materials with atoms of this configuration are more rigid, harder and stronger. Also they are less dense since the side branches prevent the chains being packed so closely together. Heat energy is required to break down the side branches and this raises their melting point above that for materials with a simple linear chain. An example of a thermoplastic material with a branched linear chain is *polypropylene*.

The cross-linked molecular chain shown in Fig. 8.6 (c) is typical of the thermosetting plastics. Thermosetting plastics are rigid and tend to be brittle once the links are formed by 'curing' the material during the moulding process. Once curing has occurred the process is not reversible and thermosetting plastics cannot be softened by reheating. If heated sufficiently they char and are destroyed.

8.4 Thermosetting plastics

Thermosetting plastics differ from the thermoplastic materials in that polymerisation (i.e. the forming of polymers, curing) is only completed during the moulding process, and from then onwards the material can never again be softened by heating.

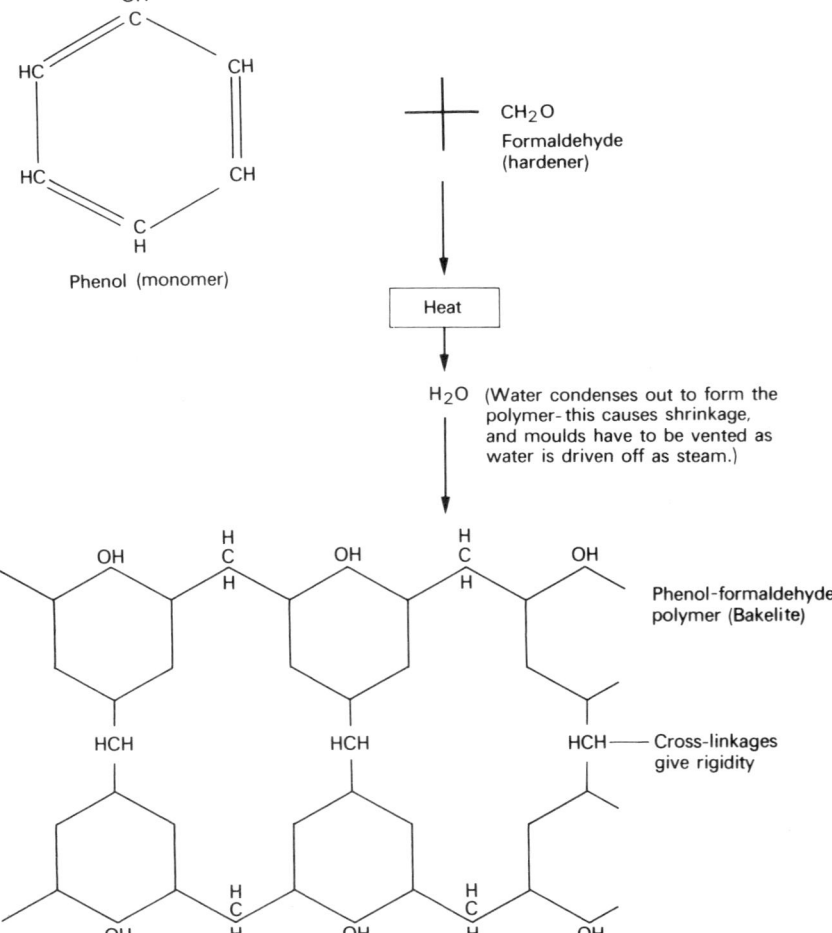

Fig. 8.7 Curing of thermosetting plastics

The difference between thermoplastics and thermosetting plastics is the way in which polymerisation occurs. In the thermoplastics discussed in sections 8.2 and 8.3 polymerisation occurs through the addition of monomers. In thermosetting plastics the polymerisation usually occurs through condensation. In this latter process the plastic material reacts with itself, or some other chemical (hardener), when heated to a critical temperature and releases or 'condenses' out some small molecule such as water. This loss of water causes shrinkage that has to be allowed for in the moulding process. Also, the moulds have to be designed with vents to allow the steam generated during the polymerisation process to escape. The principle of condensation polymerisation (curing) is shown in Fig. 8.7. Typical resins used for making thermosetting plastic moulding powders are (a) phenol formaldehyde; (b) urea formaldehyde; (c) melamine formaldehyde.

Urea formaldehyde and melamine formaldehyde are both derived from ammonia and are referred to as *amino plastics*. They do not have the strength, heat resistance and dimensional stability of phenol formaldehyde, but they have the advantage that they may be coloured white or any colour, light or dark. Phenol formaldehyde can only be coloured black or dark brown. Urea formaldehyde is the cheapest of the three thermosets being discussed, but has a low scratch resistance. Melamine formaldehyde is the hardest of all plastic materials and its resistance to scratching and scouring renders it suitable for tableware.

Thermosetting plastics are unsuitable for use by themselves, and the moulding powder used by a plastic component manufacturer contains additives to make it more economical to use and improve its mechanical properties. A typical thermosetting moulding powder could consist of:

Resin	38% by weight
Filler	58% by weight
Pigment	3% by weight
Mould release agent	0.5% by weight
Catalyst	0.3% by weight
Accelerator	0.2 by weight

The filler has a considerable influence on the properties of the mouldings produced from a given material. They improve the impact strength and reduce shrinkage during moulding. Typical fillers are:

Glass fibre — good electrical insulation properties

Wood flour
calcium carbonate } low cost, high bulk, low strength

Asbestos — heat resistance
Aluminium powder — mechanical strength

Shredded paper
shredded cloth
mica granules } good strength, combined with reasonable electrical insulation properties.

The pigment gives colour to the finished product. The mould release agent prevents the moulding from sticking to the mould. It also acts as an internal lubricant and helps the moulding material to flow during the moulding process.

The catalyst promotes the curing process during moulding.

The accelerator speeds up the curing process. Sometimes a stabiliser is added to prevent curing at room temperatures whilst not affecting the curing time at moulding temperatures.

Figure 8.8 shows the stages in manufacturing a typical thermosetting moulding material such as paper-filled melamine formaldehyde. The moulding powder or granules are delivered to the component manufacturer in paper sacks or metal drums.

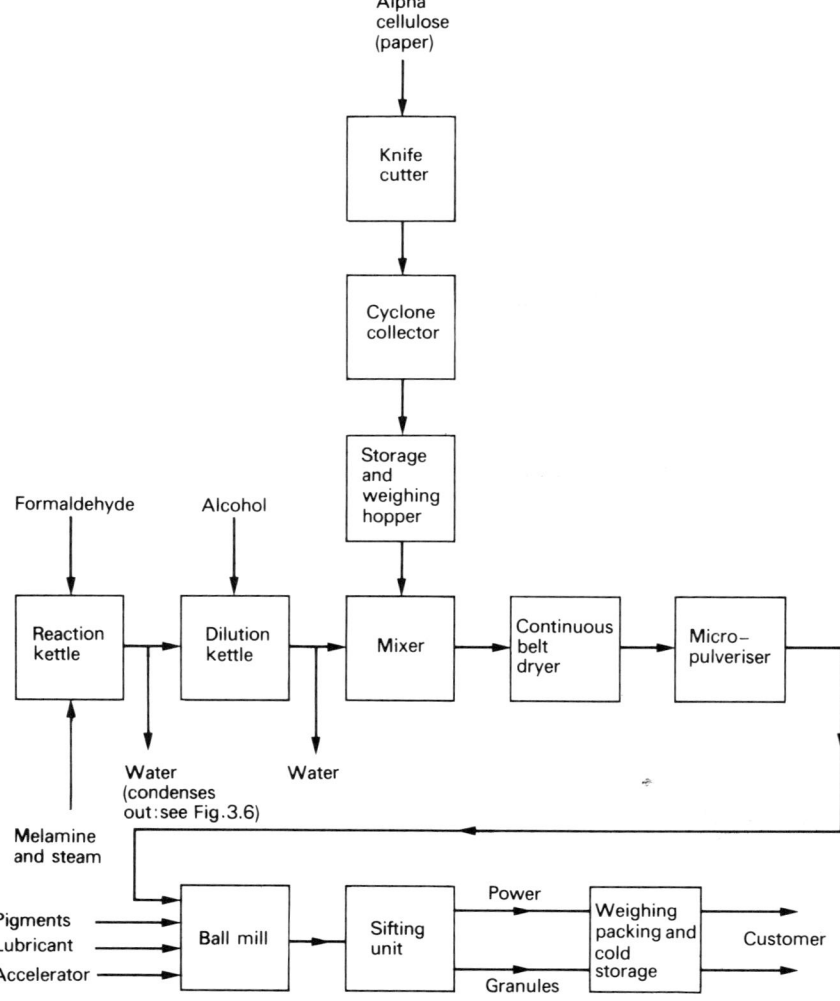

Fig. 8.8 Manufacture of a typical moulding material

8.5 Crystallinity in polymers

Crystals were described in Chapter 1 as having their particles arranged in recurring well-ordered geometric patterns. Materials which do not have this ordered arrangement of geometric patterns as their basic structure are said to be amor*phous* (without shape). For example the polymeric material PTFE has a carbon chain which describes a helix with 14 carbon atoms per turn of the helix to which are attached side chains. It is hardly surprising that a polymer with the shape of a coil spring with side chains or attached methyl groups (CH_3), or aromatic rings, has little chance of taking up the ordered patterns of a crystalline material. Hence most polymeric materials are amorphous or even glass-like (supercooled, high viscosity liquids).

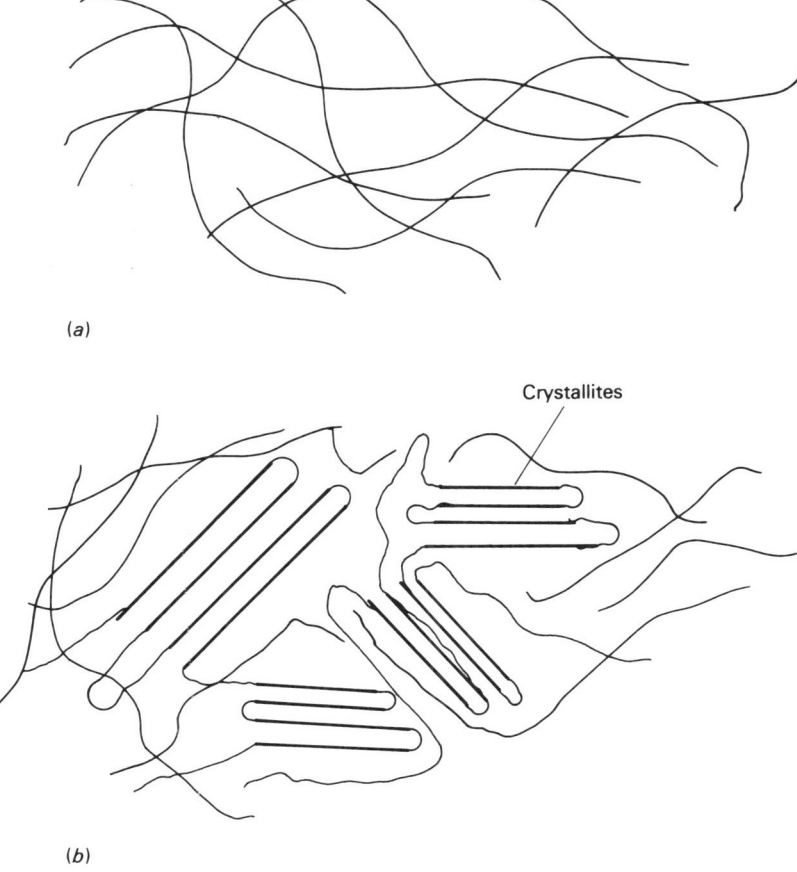

Fig. 8.9 Crystallinity in polymeric materials (*a*) Linear amorphous polymer chains (*b*) Crystallites amongst amorphous chains

Such amorphous arrangements of polymer chains are often indicated by a tangle of lines as shown in Fig. 8.9 (*a*). Each line represents an individual molecular chain, but does not show the individual atoms for the practical reason that there may be many thousands of atoms spread along each chain. However, simple linear chains without side branches or cross-links may show some degree of ordering on a sub-microscopic scale. Such ordered regions are called *crystallites* and there may be several such regions along a single molecular chain. Such an arrangement is shown in Fig. 8.9 (*b*) where it can be seen that the individual molecular chains extend through several crystalline and non-crystalline regions.

The *crystallinity* of a polymeric material is defined as the ratio between the mass of the crystallites and the total mass of the material being considered. For example a material having 80 per cent crystallinity will consist of 80 per cent crystallite structure and 20 per cent non-crystallite or amorphous structure. Since the monomers making up a polymer chain are packed more closely together in crystallites, it follows that materials with a high crystallinity will be more dense than materials with a low crystallinity. For example, low-density polyethylene with a crystallinity of only 50 to 70 per cent will have a density of about 920 kg m^{-3} and a melting point of 115 °C; whereas high-density polyethylene with 75 to 95 per cent crystallinity will have a density of about 950 kg m^{-3} and a melting point of 135 °C.

The crystallinity of a polymeric material has a marked effect upon its properties. For example, increasing the crystallinity of a material:

1. increases its melting point and, instead of softening gradually with increased temperature, it will exhibit a sharper melting point which is characteristic of most crystalline materials;
2. increases the resistance of the material to the absorption of water and to solvent attack since it is more difficult for the water and solvent to penetrate the high-density crystallites than it is to penetrate the more open amorphous structure;
3. prevents the penetration of plasticisers and this reduces the ultimate elongation of the material, (Fig. 8.10);
4. makes the material more impervious to gasses and this may be useful in food packaging and protective coatings. However, this high level of impermeability is a disadvantage in polymer fibres which must be coloured by dyeing.

The effect of crystallinity on the ultimate tensile strength and percentage elongation of a typical polymeric material such as polyethylene is shown in Fig. 8.10. The relative crystallinity can be altered by heat treatment. A crystalliable polyethylene can give a crystallinity of 80 per cent by slow cooling or only 65 per cent by rapid cooling (quenching).

The intermediate condition of *orientation*, which lies between the amorphous and crystalline state, will be considered in *Materials technology: level 3*.

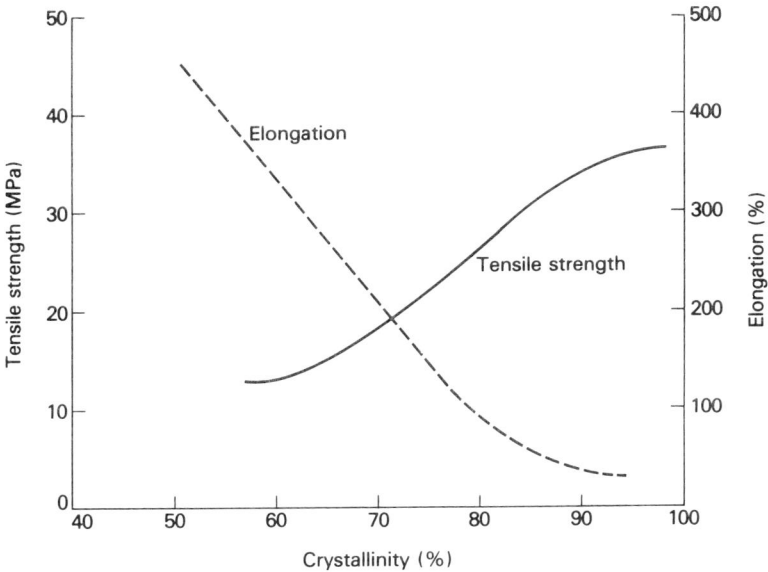

Fig. 8.10 Effect of crystallinity on the properties of polyethylene

8.6 General properties of polymeric materials

The properties of plastics can vary widely, but all plastic materials have the following properties in common:

Electrical insulation
All polymeric materials exhibit good electrical insulation properties. However, their usefulness in this field is limited by their low heat resistance and softness. Thus they are useless as formers on which to wind electric radiator elements, and as insulators for use out of doors where their surface would soon be roughened by the weather. Dirt collecting on this roughened surface would then provide a conductive path, causing a short circuit.

Strength/weight ratio
Polymeric materials vary in strength considerably. Some of the stronger, such as nylon, compare favourably with the weaker metals. All of them are much lighter than most metals. Therefore, properly chosen and proportioned, their strength/weight ratio compares favourably with many light alloys. They are steadily taking over engineering duties which, until recently, were considered the prerogative of metal. The tensile strength of typical thermoplastics is shown in Fig. 8.11.

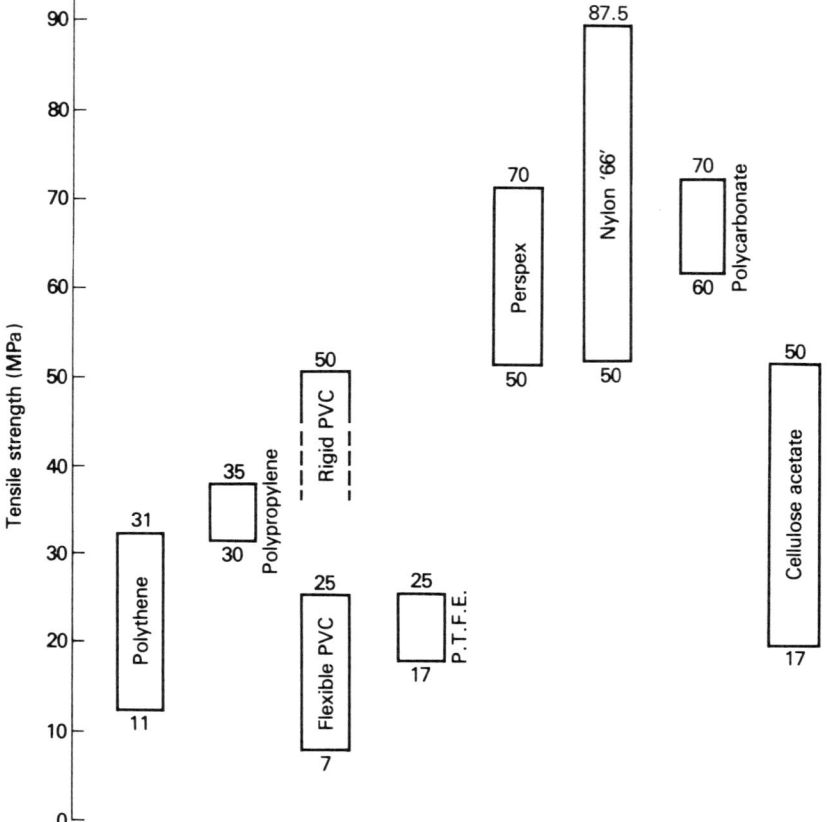

Fig. 8.11 Tensile strengths of some typical thermoplastic groups

Corrosion resistance

All polymeric materials are inert to most inorganic chemicals. Thus they can be used in environments that are hostile to the most corrosion-resistant metals.

They are superior to rubber in that they are resistant to attack by oils and greases.

8.7 Specific properties (polyethylene)

This thermoplastic material, which is also known as *polythene*, was considered in section 8.6 as a typical example of a polymeric material which can exist in the amorphous state or with varying degrees of crystallinity. Low density polyethylene has a branched molecular chain which makes the formation of crystallites difficult. The structure of the branched molecular chain is shown in Fig. 8.12 (*a*). High density polyethylene has a simple linear molecular chain and this lends itself to the close packing essential to the orderly structure required to form crystallites. The structure of a simple linear chain of polyethylene is shown in Fig. 8.12 (b). Table 8.3 compares the properties of typical high- and low-density polyethylenes.

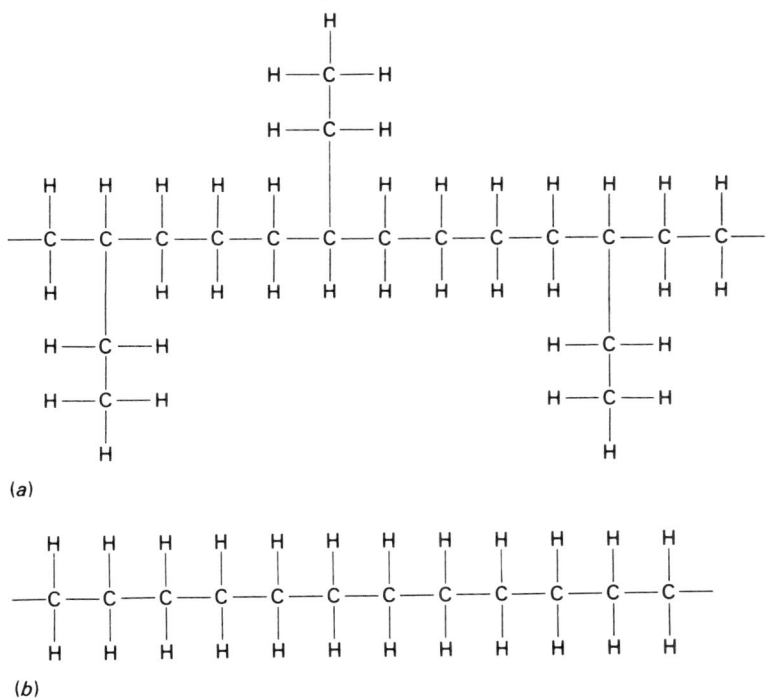

Fig. 8.12 Polymer chains for polyethylene (a) Polymer chain for low-density (branched) polyethylene (b) Polymer chain for high-density (linear) polyethylene

Table 8.3 Properties of polyethylene

Material	Properties				
	Density (kg m^{-3})	Melting point	Tensile strength (MPa)	Elongation (%)	Maximum service temperature °C
High-density Polyethylene	950	135	22–40	50–800	125
Low-density Polyethylene	920	115	8–15	100–600	85

Low-density polyethylene softens in boiling water and melts at only just above this temperature. It is used for films and sheeting for agricultural, horticultural and building purposes. It is made into bags for food storage, flexible tubing and electric wire and cable insulation.

High-density polyethylene is used for piping which has to carry hot water, toys, fibres for fabrics and household articles such as washing-up bowls. Both high- and low-density polyethylene have good resistance to chemical attack, low moisture absorption, and high electrical resistance.

The properties of high- and low-density polyethylene can be enhanced by:

1. blending them to give intermediate properties;
2. adding carbon black as a stabiliser and to retard the degradation caused by strong sunlight;
3. pigments to give colour and make the material more attractive;
4. reinforcement with fibrous materials;
5. adding butyl rubber to maintain flexibility and prevent in-service cracking.

8.8 Specific properties (polyamides)

Polyamides are that group of polymeric materials known as *nylons* and were the first of the 'engineering' or high-strength thermoplastics. Polyamides are produced by the reaction of diamine with an organic acid. One of the most commonly used nylons is produced by reacting hexamethylenediamine with adipic acid to give hexamethylenedipamide (commonly polyamide) which is called nylon 6/6. This notation indicates that there are six carbon atoms in the amine and six in the acid segments. Other grades of nylon are 6/10, 6/12, 13/13, etc. The molecular structure of nylon 6/6 is shown in Fig. 8.13. This is just one segment from the molecular chain. Since the molecular chain is linear, it gives rise to crystalline structures. Nylon materials are tough and stiff. Nylon has a low coefficient of friction and good abrasion resistance. This results in its being widely used for light duty, moulded gears, bushes and cams. It has the advantage for office machinery and food processing machinery of not requiring lubricating. Unfortunately nylons absorb water, even from the atmosphere, and this makes them poor electrical insulators and results in lack of dimensional stability.

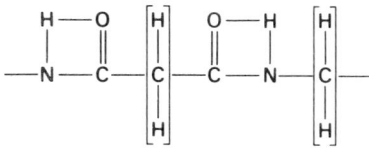

Fig. 8.13 Segment from nylon '66' polymer chain

This absorbed moisture acts as a plasticiser, lowering the stiffness, strength and hardness. The effect on their strength is shown in Fig. 8.14. Table 8.4 gives the average properties that may be expected from some typical nylons. As well as being moulded to shape nylon is also extruded as fibres for making such diverse articles as clothing and high-tensile ropes.

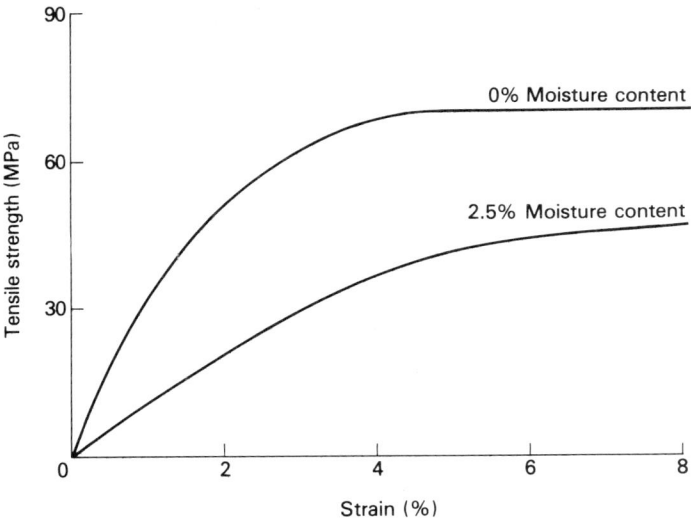

Fig. 8.14 Effect of moisture content on the strength of nylon '66'

Table 8.4 Properties of typical nylons

Material	Properties			
	Density (kg m^{-3})	Tensile strength (MPa)	Elongation (%)	Maximum service temperature (°C)
Nylon 6	1100	70–90	60–300	120
Nylon 6.6	1100	80	60–300	120
Nylon 6.10	1100	60	80–230	120
Nylon 11	1100	50	75–300	120

8.9 Polyvinyl chloride

Polyethylene consists of carbon and hydrogen atoms arranged in long-chain molecules. Since both carbon and hydrogen are flammable it follows that polyethylene is also highly flammable. This makes it unsuitable for many purposes where such flammability would contravene building and insurance regulations. However, by replacing a hydrogen atom with a chlorine atom in each monomer present in the molecular chain, a non-flammable material is produced (see Fig. 8.5). This process is called chlorination and the modified monomer is called polyvinyl chloride (PVC). Polyvinyl chloride can be made rigid or flexible by adding a plasticiser. The effect of a plasticiser on polyvinyl chloride is shown in Table 8.5.

In its rigid form polyvinyl chloride is widely used for rain guttering and fittings, waste pipes for sinks and wash basins. It is not suitable for hot water as its service temperature is only 70 °C. Plasticised polyvinyl chloride is used for waterproof clothing, bottles, shoe soles, garden hose, toys, etc. With the aid of suitable additives it can withstand prolonged exposure to strong sunlight with no appreciable degradation.

Table 8.5 Effect of plasticiser content on polyvinyl chloride

Plasticiser Content	Properties			
	Density ($kg\ m^{-3}$)	Tensile Strength (MPa)	Elongation (%)	Maximum Service Temperature (°C)
Nil	1400	50–58	2–40	70
Low	1300	30–40	200–250	60–100
High	1200	15–20	350–450	60–100

8.10 Polystyrene

Like polyethylene and polyvinyl chloride, polystyrene is a low-cost, widely used thermoplastic material. From Fig. 8.15 it can be seen that it has an aromatic ring as a side branch to the monomer. A long chain molecule made from monomers will be too bulky to pack closely together to form crystallites, thus polystyrene will have an amorphous structure. Two general grades of polystyrene are available. These are general-purpose polystyrene and high-impact polystyrene.

Amorphous polystyrene is a crystal clear material which is hard, brittle and low in impact resistance. It is easily stress-cracked and has a maximum service temperature of only 65 °C. It is not of much use

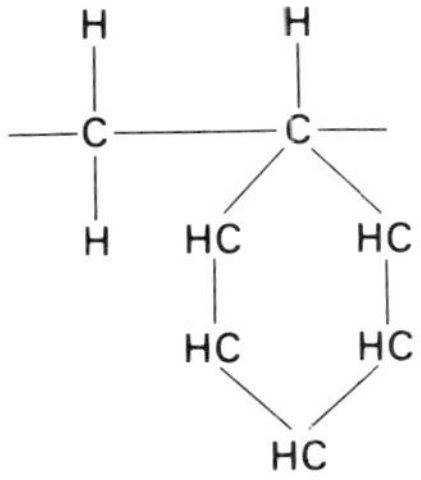

[Structural diagram of polystyrene segment]

Fig. 8.15 Segment of the polymer chain for polystyrene

commercially. By rearranging the side branches more uniformly some degree of crystallinity can be achieved and this greatly improves the properties of polystyrene. This crystalline polystyrene is referred to as general-purpose polystyrene and is used for packaging cosmetics, and making toys, decorative light-fittings, drinking tumblers, etc.

High impact polystyrene is made by copolymerising it with about 5 per cent butadiene synthetic rubber. This material is far less brittle and is used for casings for cameras, projectors, radios, etc., for domestic use.

Another variant is to copolymer high-impact polystyrene with acrylonitrile to form chains of three polymer materials as shown in Fig. 8.16. This material is called acrylonitride-butadiene-styrene terpolymer (happily abbreviated ABS). This material is tough, stiff and abrasion resistant. It has a wide range of applications from casings for telephones, hair driers, and radios, to safety helmets and boat hulls. The properties of these three variants on styrene polymers are compared in Table 8.6.

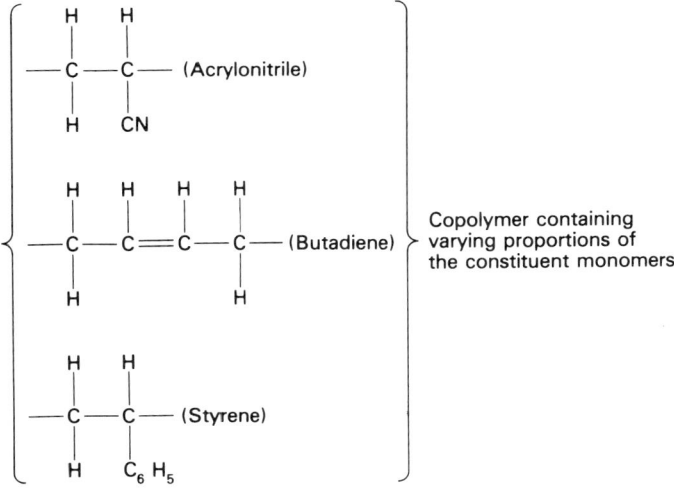

Fig. 8.16 Segment of the copolymer of ABS

Table 8.6 Properties of styrene group thermoplastics

Materials	Density (kg m^{-3})	Tensile strength (MPa)	Elongation (%)	Maximum service Temperature (°C)
Low-impact Polystyrene	1100	35–60	1–5	65
High-impact (toughened) Polystyrene	1100	15–40	8–50	75
ABS	1100	18–60	10–140	110

Another widely used form of polystyrene is *expanded polystyrene*. This is a rigid cellular structure containing gas-filled bubbles. Since gas is a poor thermal conductor and since it is entrapped in the individual bubbles, expanded polystyrene is widely used for thermal insulation. Because of the exceptionally low density of this rigid foam material and the ease with which it can be moulded, it is widely used as a protective material in packaging.

No examples of a thermoset have been included. This is because thermosetting resins are always blended with such a high percentage of filler materials that the filler has as much influence on the overall properties of the moulding as the resin itself. See section 8.4.

8.11 Elastomers

The elastomers, or rubbers, are cross-linked polymeric materials. However, there are not sufficient cross-links to make them as rigid as the thermosetting plastics, but just sufficient to make them return to their original dimensions when the deforming load is removed. Whereas thermosets show little elongation under stress, elastomers are capable of elongations of up to 1000 per cent at tensile failure. Elastomers are, therefore, capable of extreme elastic deformation at low levels of stress. The strain is not proportional to stress and this is shown in Fig. 8.17. This is the typical S-curve exhibited by elastomers.

The elastomers are usually addition polymerised as thermoplastics and then cross-linked (vulcanised) with sulphur at approximately every five-hundredth carbon atom. Increased vulcanisation increases the cross linking and this, in turn, increases the stiffness and reduces the elongation percentage of the material. Fully vulcanised, natural rubber becomes a rigid, brittle thermoset called ebonite. Some typical elastomers are as follows:

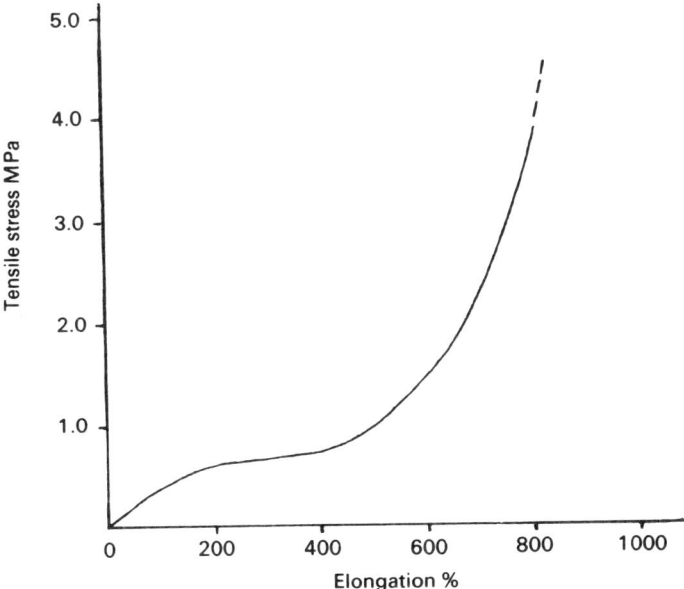

Fig. 8.17 Typical stress-elongation curve for a polyisobutylene rubber

1. Styrene-butadiene rubber (SBR) A general purpose synthetic rubber used for tyres, belts, floor tiles and latex paints. It is superior to natural rubber in respect of skid resistance, solvent resistance and weathering.

2. Polyisoprene rubber (natural rubber) This is derived from the sap of a tree called *Hevea brasiliensis*. It has a low hysteresis and a high tensile strength. Unfortunately it is readily attacked by solvents, mineral oils, ozone and petrol. It degrades (perishes) in the presence of strong sunlight.

3. Butyl rubber This rubber is impervious to gases and is used as a vapour barrier and hose lining. It is highly resistant to outdoor weathering and ultraviolet radiation.

4. Nitrile rubber This has excellent resistance to oils and solvents, and can be readily bonded to metals. It is used for petrol hose and hose linings, aircraft fuel tank linings. It is also resistant to refrigerant gases.

5. Polychloroprene rubber (neoprene) This was the original synthetic rubber developed during World War II. It has good resistance to oxidation, ageing, and weathering. It is resistant to oils and solvents, abrasion and elevated temperatures. Because of its chlorine content it is fire resistant. It is used as a flexible electrical insulator and for gaskets, hoses, engine mounts, sealants, rubber cements and protective clothing.

6. Polysulphide rubber (thiokol) Although this rubber has low mechanical strength, its resistance to solvents and its impermeability to gases is excellent. Its weathering characteristics are outstanding. It also has good bonding properties and is widely used in the construction industries as a sealant. Thiokol and polyurethane rubbers are also used as fuels for solid-fuel rockets.

7. Acrylic rubbers These are derived from the same family of polymeric materials as *perspex* but not so heavily cross-linked to render them rigid. They have excellent resistance to oils, oxygen, ozone and ultra-violet radiation. They are used as a basis for the latex paints used on motor vehicles.

8. Rubber hydrochloride This material is better known as *Pliofilm* and is used to form a transparent film for the vacuum packaging of foodstuffs and DIY hardware. It is easily identified by its unusual tensile and tear strength.

9. Silicone rubbers Although silicone rubber has a relatively low tensile strength, it has an exceptionally wide working temperature range of $-80\,°C$ to $+235\,°C$, thus often outperforming other rubbers which, superior at room temperatures, cannot exist at such temperature extremes. It can be used for mould linings and high-temperature seals.

10. Polyurethane rubber Polyurethane can be formulated to give either plastic or elastomer characteristics. Although it has high strength and abrasion resistance, it is of little use as a tyre material as it has a low skid resistance. However, its outstanding service life makes it suitable for solid cushion tyres for warehouse trucks, fork-lift trucks and similar warehouse vehicles where low speeds and dry floor conditions do not make its low skid resistance so important. It is also used for shoe heels, painting rollers, mallet heads, oil seals, diaphragms, anti-vibration mountings, gears, pump impellers, etc.

The uses to which elastomers (rubbers) may be put in engineering may be classified as follows:

1. Vibration insulation and isolation
   (*a*) Shock absorbers.
   (*b*) Anti-vibration machine mountings.
   (*c*) Sound insulation.

2. *Distortional systems*
   (a) Correctives for misalignment such as flexible couplings.
   (b) Changing shapes such as belts, flexible hose, covered rolls, tyres, etc.
   (c) Seals of all kinds and gaskets.
   (d) Rubber hydraulics.

3. *Protective systems*
   (a) Protection against abrasion.
   (b) Protection against corrosion.
   (c) Electrical insulation.

Problems

Section A

1. State the main difference between thermoplastic materials and thermosetting plastic materials.
2. Name the four groups of hydrocarbons which are used as the basic feed stocks for polymeric materials.
3. Describe the difference between crystalline and amorphous polymeric materials and how this affects their properties.
4. Select a suitable polymeric material for the following applications giving reasons for your choice: (i) electric cable insulation; (ii) non-stick coating of cooking utensils;   (iii) high strength, lightweight ropes for mountaineering;   (iv) low friction, oil-less bushes for office machinery;   (v) heat insulation blocks.
5. What is meant by the term an 'elastomer'.

Section B

6. Describe in detail the differences between the following groups of substances and explain how they affect the polymeric materials made from them: (i) paraffins;   (ii) olefins; (iii) naphthenes;   (iv) aromatics.
7. (a) Explain the reasons for the inclusion of the following additives in a thermosetting moulding powder: (i) resin; (ii) filler;   (iii) pigment;   (iv) mould release agent;   (v) catalyst;   (vi) accelerator.
   (b) State typical applications for the following filler materials: (i) glass fibre;   (ii) wood flour; (iii) calcium carbonate;   (iv) aluminium powder; (v) shredded paper;   (vi) mica granules.
8. (a) With the aid of diagrams explain what is meant by the following terms relating to polymer chains:
   (i) linear;   (ii) branching;   (iii) cross-linked.
   (b) With the aid of a diagram show how a typical thermosetting resin is cured (polymerised) by the condensation of water molecules. What precautions have to be taken in the design of the moulds to compensate for the curing process?

9. (a) With the aid of diagrams describe how crystallinity occurs in a polymeric material and explain in detail how this affects the properties of such materials.
   (b) Contrast and compare the general properties of polymeric materials with those of the more common metals.
10. Select a typical example of each of the following groups of polymeric materials, and describe its properties and typical applications: (i) thermoplastic;   (ii) thermosetting plastic;   (iii) reinforced plastic;   (iv) elastomer; (v) rubber.

Index

Acrylic rubber, 147
Acrylics, 167
Allotropy, 10
Alloy types, 16, 17, 24
Alloys – cast iron, 109, 112, 113
Aluminium alloys, 119 et seq.
Aluminium bronze alloys see copper alloys
Aluminium casting alloys
   (non-heat-treatable), 119, 120, 122
   (heat-treatable), 124, 125, 126
Aluminium
   commercial, purity, 115
   high purity, 115
   properties of, 118, 119
Aluminium wrought alloys
   (non-heat-treatable), 121, 122, 124
   (heat-treatable), 126, 127
Amino resins, 146
Annealing
   full, 60
   process, 57 et seq.
   spheroidising, 58–60
   stress relief, 58
Aromatics, 150–1
Atmosphere control, 76, 93, 94, 95, 96
Atoms, 1, 3, 4
Austenite, 39

Bainite, 64
Bakelite see phenolic resins
Binary equilibrium diagrams, 16 et seq.
Brass see copper alloys
Butyl rubber, 167

Carburising
   gas, 70–1
   heat treatment after, 71–2
   pack, 70
   salt bath, 70
Case hardening, 68–70
   localised, 72–4
Cast irons, 100 et seq.
   alloying elements and impurities of, 103–4
   iron-carbon system for, 100–2
   manganese, effect of, 103
   phosphorus, effect of, 104
   properties and uses of, 109–12
   silicon, effect of, 103
   sulphur, effect of, 103
Cellulose plastics, 147
Cementite, 40
Cold working, 57

Compounds, 7
Cooling curves, 21–4
Copper, 126, 128–30
Copper alloys
   aluminium bronze, 137–8, 140
   beryllium copper, 131
   brass, 132–4
   cadmium copper, 130–1
   chromium copper, 131
   cupro-nickel, 139, 141
   silver copper, 130
   tellurium copper, 131
   tin-bronze, 135–7
Coring, 31–3
Critical change points, 44–5
Critical temperatures, 57
Crystal growth, 11–14
Crystallinity see polymers
Crystals, 7
Cupro-nickel alloys see copper alloys

Elastomers, 166–8
Electrons, 4
Elements, 5–6
End quench test see Jominy Test
Energy sources, 86–7
Epoxy resin, 146
Eutectoid point, 40

Ferrite, 39
Ferrous metals, 36 et seq.
Fuel – economical use of, 76, 77
Furnace
   double chamber, 82–3
   muffle (electrically heated), 80–1
   muffle (gas heated), 79–80
   open hearth, 77–8
   requirements of a heat treatment, 75–7
   salt bath (electrically heated), 85, 86
   salt bath (gas-fired), 83–5
   semi-muffle, 78–9

Grain structure, 10–11
Grey cast-iron, 101, 102
   heat treatment of, 104–5

Hardenability, 67–8
Hardening (quench), 62–4
Heat treatment
   causes of cracking and distortion, 98
   processes, 55 et seq.
Hot working, 57

Intermetallic compounds, 20
Ions, 4
Iron-carbon system, the, 37–44
Isotopes, 4

Jominy end quench test, 68

Magnesium alloys, 139, 142
Malleable Cast Iron
  (Blackheart process), 103, 105
  (Pearlitic process), 107, 109
  (Whiteheart process), 105, 107
Martensite, 64
Mass effect, 66–8
Metals
  ferrous, 36 et seq.
  non-ferrous see Non-ferrous metals and alloys
  structure of, 1 et seq.
Mixtures, 6–7
Molecules, 5

Naphthenes, the, 150–1
Nitrile rubber, 167
Non-ferrous metals and alloys, 114–15, 116–17
Normalising, 60–1
Nucleus, 4

Olefins, the, 149

Paraffins, the, 149
Pearlite, 40, 41, 42
Phenolic resins (Bakelite), 146
Plain carbon steels, 48 et seq.
  effect of carbon on properties of, 45–7
  heat treatment of, 55 et seq.
Plastics
  building blocks of, 146, 147, 148
  glass reinforced (GRP), 146
  laminated, 145
  thermo-, 145, 147
  thermosetting-, 145, 146, 154, 155–6
Polyamides, 147
  specific properties of, 162–3
Polychloroprene (neoprene) rubber, 167
Polyesters, 147
Polyethylene, 147
  specific properties of, 160–2
Polyisoprene (natural) rubber, 167
Polymeric materials, general properties of, 159–60
Polymers, 152–4
  crystallinity in, 157–8
Polypropylene, 147
Polystyrene, 147
  specific properties of, 164–5

Polysulphide rubber, 167
Polytetrafluoroethylene, 147
Polythene, 147
Polyurethane rubber, 168
Polyvinyl chloride, 147
  specific properties of, 164
Precipitation, 33–4
  hardening, 34
Protons, 4
Pyrometer
  optical, 92
  radiation, 91–2
  thermo-couple, 88–91

Quench hardening see hardening
Quenching media, 93, 97

Recrystallisation, 55–7
Rubber see elastomer
Rubber hydrochloride (pliofilm), 167

Safety (heat treatment), 97, 99
Silicone rubber, 168
Solid solutions, 18–20
Solubility, 17–18
Solution treatment (aluminium alloy), 34
Sorbite, 64
Spheroidal graphite (S.G.) cast Iron, 108, 109
Steels
  hyper-eutectoid, 40
  hypo-eutectoid, 40
  plain carbon, 36, 48–53
Styrene – butadiene rubber (SBR), 167

Temperature measurement, 87 et seq.
Tempering, 65, 66
Thermal equilibrium diagram
  combination type, 28, 29, 30, 31
  eutectic type, 24–5, 26, 27
  solid solution type, 27–8
Thermometer
  mercury in steel, 88
  vapour pressure, 88
Thermoplastic see plastics
Thermosetting plastic see plastics
Tin-bronze alloy see copper alloys
Troostite, 64

Vinyl plastics, 147

Wrought iron, 47

Zinc alloys, 143